THE EVOLVING BOUNDARIES OF DEFENCE: AN ASSESSMENT OF RECENT SHIFTS IN DEFENCE ACTIVITIES

Contributions to

Conflict Management, Peace Economics and Development Volume 23

SERIES EDITOR

MANAS CHATTERJI

BOOKS IN THE SERIES

Eurasia: A New Peace Agenda, edited by M. D. Intriligator, A. I. Nikitin and M. Tehranian

Military Missions and Their Implications Reconsidered: The Aftermath of September 11th, edited by G. Caforio and G. Kümmel

Managing Conflict in Economic Convergence of Regions in Greater Europe, by F. Carluer

Cultural Differences between the Military and Parent Society in Democratic Countries, edited by G. Caforio

Conflict and Peace in South Asia, edited by M. Chatterji and B. M. Jain

War, Peace, and Security, edited by J. Fontanel and M. Chatterji

Armed Forces and Conflict Resolution, edited by G. Caforio, G. Kümmel and B. Purkayastha

Regional Development and Conflict Management: A Case for Brazil, by R. Bar-El

Crisis, Complexity and Conflict, by I. J. Azis

Putting Teeth in the Tiger: Improving the Effectiveness of Arms Embargoes, edited by M. Brzoska and G. A. Lopez

Peace Science: Theory and Cases, by P. Gangopadhyay and M. Chatterji

Advances in Military Sociology: Essays in Honor of Charles C. Moskos (Two Volume Set), edited by Giuseppe Caforio

Arms and Conflict in the Middle East, by Riad A. Attar

Economics of War and Peace: Economic, Legal, and Political Perspectives, edited by Benjamin E. Goldsmith and Jurgen Brauer

Conflict, Complexity and Mathematical Social Science, by Gordon Burt

Frontiers of Peace Economics and Peace Science, edited by Manas Chatterji, Chen Bo and Rameshwar Misra

Ethnic Conflict, Civil War and Cost of Conflict, edited by Raul Caruso

Governance, Development and Conflict, edited by Manas Chatterji, Darvesh Gopal and Savita Singh

New Wars, New Militaries, New Soldiers? Conflicts, the Armed Forces and the Soldierly Subject, edited by Gerhard Kümmel and Joseph Soeters

Cooperation for a Peaceful and Sustainable World, Part 1, edited by Chen Bo, Manas Chatterji and Hao Chaoyan

Cooperation for a Peaceful and Sustainable World, Part 2, edited by Li Junsheng, Chen Bo and Hou Na

Nuclear Disarmament: Regional Perspectives on Progress, edited by P. M. Kamath

Understanding Terrorism: A Socio-Economic Perspective, edited by Raul Caruso and Andrea Locatelli

Contributions to
CONFLICT MANAGEMENT, PEACE ECONOMICS AND DEVELOPMENT VOLUME 23

THE EVOLVING BOUNDARIES OF DEFENCE: AN ASSESSMENT OF RECENT SHIFTS IN DEFENCE ACTIVITIES

EDITED BY

RENAUD BELLAIS

Pôle Sciences Humaines et Sociales, ENSTA Bretagne, Brest, France

Emerald

United Kingdom – North America – Japan
India – Malaysia – China

Emerald Group Publishing Limited
Howard House, Wagon Lane, Bingley BD16 1WA, UK

First edition 2014

British Library Cataloguing in Publication Data
A catalogue record for this book is available from the British Library

ISBN: 978-1-78350-974-4
ISSN: 1572-8323 (Series)

CONTENTS

LIST OF CONTRIBUTORS

Jacques Aben	Centre de Recherches Interdisciplinaires en Sciences Humaines et Sociales (CRISES), Université de Montpellier, Montpellier, France
Francesca Artioli	Centre d'Études Européennes (CEE), Sciences Po Paris, Paris, France
Renaud Bellais	Pôle Sciences Humaines et Sociales, ENSTA Bretagne, Brest, France
Steven Coissard	Research Department, IDRAC Business School, Lyon, France
Bénédicte Corvaisier-Drouart	Institut d'Études Politiques de Grenoble, Grenoble, France
Roland de Penanros	Department of Economics, Université de Bretagne Occidentale, Brest, France
Josselin Droff	Department of Economics, Université de Bretagne Occidentale, Brest, France and Pôle Sciences Humaines et Sociales, ENSTA Bretagne, Brest, France
Jacques Fontanel	Department of Economics, Université Pierre Mendès France, Grenoble, France
Gueorgui Ianakiev	DG Competition, European Commission, Brussels, Belgium
Susan Jackson	Department of Global Political Studies, Malmö University, Malmö, Sweden
Tanu Kohli	College of Liberal Arts & Sciences, Program in Global Studies, University of Illinois at Urbana-Champaign, Champaign, IL, USA

Scott Yi-Chun Lin	Graduate Institute of Development Studies, National Chengchi University, Taipei City, Taiwan
Alfredo R. Paloyo	Centre for Human and Social Capital Research, University of Wollongong, Wollongong, Australia
Liliane Perrin-Bensahel	Institut d'Urbanisme de Grenoble, Université Pierre Mendès France, Grenoble, France
Thierry Sauvin	Department of Economics, Université de Bretagne Occidentale, Brest, France
Carlos Seiglie	Department of Economics, Rutgers University, Newark, NJ, USA
Cindy Williams	Security Studies Program, Center for International Studies, Massachusetts Institute of Technology, Cambridge, MA, USA
Rachel Woodward	School of Geography, Politics and Sociology, Newcastle University, Newcastle upon Tyne, UK

FOREWORD

Military spending and defense spending are not synonymous. They do not necessarily increase security. The relationship of military spending and economic development has been researched for more than thirty years. No definitive conclusion has been reached. Due to the end of cold war and onset of globalization, the role of military has changed drastically. The world is in a post-interventionist era and sanctions and diplomacy are replacing military intervention. Technology has also changed traditional role of military. Economic conversion from military to civil production is a matter of great importance, particularly in Russia and Eastern Europe. Military contract and its multiplier impact are also becoming more crucial for the regional economy and employment.

During the cold war, a significant portion of economic resources used to be devoted to military spending. After the cold war is over, it was thought there will be some Peace Dividend. That did not take place. Military expenditure increased steadily in the beginning of this century primarily due to the global war on terrorism. Peace time spending was about the same to that of cold war times. Another characteristic of military spending is its spatial character, that is, how defense expenditure is spread over geographical space and is inter-regionally connected. Defense is also based on the cooperation of different countries like NATO. The nature of deployment of military units depends on this coalition. Dichotomy between defense and non-defense matters is fast disappearing. The cost of defense will increase substantially in the future due to the nature of the weapon system.

In sociological literature, there are a number of studies about military and civilian culture, including some cross-cultural research about the civil and military societies. Human resources involving such factors as gender have also changed this situation particularly due to the abolition of draft. Military is a top-down organization. It is more centralized and formalized. Sociologists predict a gap between organizational culture of military and civilian societies. The gap between US military and civilian society is growing very fast. The leadership in military is skeptical about the civilian leadership. Its hostility toward media, more trust, and confidence on civilians rather than government institutions is well known.

This volume includes all these aspects in the contributions from well-respected scholars mostly from Europe.

Manas Chatterji
Binghamton University, NY, US
Series Editor

INTRODUCTION

Defence has always been a difficult topic for economics as well as many social sciences. This field of research is marked by many political interferences, which are tough to sort out and therefore lead many academics to avoid defence matters. Additionally, understanding defence issues requires most of time to mobilise many dimensions at the same time and several approaches from social sciences. This is the reason why defence has been neglected despite many pending questions are treated as a specific case diverting from basic models that cannot be approached without risks. One can thus consider that defence constitutes a 'realm' that should be analysed per se or, at least, by using caveats.

Nevertheless, there are several economic and social stakes linked to defence, and this field should not be avoided or left uncharted. Even though defence does have specific features, this does not mean that this field should not be explored: this is in fact a good incentive to understand its functioning and define its core specificities. In fact, defence takes a large place in most countries and one should wonder what impacts related activities have on the whole society. Applying economic and social approaches to defence topics is therefore necessary to provide decision-makers and citizens the right means to understand what is at stake.

The most obvious stake results from massive budgetary efforts dedicated to defence. According to the Stockholm International Peace Research Institute, world military expenditure was estimated to reach $1,747 billion in 2013, representing 2.4 per cent of global gross domestic product or $247 for each person in the world. This level of effort has impacts on the short term (warfare vs. welfare), but it takes place on a long-term trend that induces long-term economic and social trends over decades with both negative and positive impacts. Such huge spending is bound to raise many questions and this is the reason why defence cannot remain a terra incognita.

This macroeconomic dimension represents only one angle to deal with defence matters. Indeed, defence has several interactions with the whole economy and society, both domestically and internationally. This is the reason why analysing the economic impacts of defence activities requires to mobilise many approaches within economics and beyond, particularly in political science, geography or sociology. In this sense too, defence constitutes a 'realm' to be explored and mapped in terms of research agenda.

Indeed, it would be unfair to consider that defence represents uncharted waters. Economics and social sciences have been investigating defence for decades, but mainly because of high levels of military expenditures throughout the Cold War. One could refer to the numerous books edited by Manas Chatterji, Keith Hartley or Todd Sandler. These research projects delivered many interesting and fruitful insights, with substantial advancement in the understanding in many defence-related issues. Nevertheless, many dimensions of defence activities or their impacts are still under-explored or neglected today. This is particularly true concerning trans-disciplinary issues. It appears that we are just at the edge of such an exploration.

Moreover, defence has been experiencing deep transformations in many countries because of geostrategic and economic changes since the 1990s. This constitutes a strong incentive to have a renewed look at the realm of defence in the early twenty-first century.

During the Cold War, a large share of GDP was dedicated to defence expenditures. Military activities used to absorb a significant share of resources and space in most countries. The quest for 'peace dividends' resulted in a major contraction of military spending in the 1990s, inducing in a profound redrawing of the place taken by armed forces, arms industry, military bases... in the economic and social landscape. The end of the Cold War has truly modified the place of defence matters and military-civilian relations in many countries.

One could consider that this post-Cold War period was an abnormal situation, a deviation from the traditional role and place that defence holds historically in many countries. Indeed military expenditure has increased steadily in the 2000s and the 2010s, driven by what the United States labelled as 'the global war on terrorism' and then by the expending purchasing power of emerging economies. It is true that the global military expenditure reaches an extraordinary level in the beginning of the twenty-first century, especially regarding a period that can be considered as relatively peaceful in a historical perspective. Military expenditure is indeed similar to the highest levels of the Cold War. One could then easily compare both periods in terms of theoretical and empirical treatment.

Nevertheless a similar level of military expenditure does not mean that the world came back to the ex-ante situation. On the contrary, many things have changed in the realm of defence and it seems difficult to simply compare today's defence with what took place before 1990. One cannot 'read' or understand today's defence issues only by relying on yesterday's analytical tools. It is necessary to look specifically at this new, evolved realm of defence to understand the effective stakes and avoid misinterpreting its social and economic footprints.

Indeed, there are many differences between the way we perceive the place of defence in today's world and the place it had until the 1980s. The most visible transformation can be perceived through the spatial reorganization of defence activities, both domestically in terms of military bases and defence industry and internationally in terms of deployments. Many bases have been closed, particularly in countries where armed forces have gone through a massive contraction and where draft was suspended to move to all-volunteer forces. Similarly, today's missions and operations present several features that distinguish them from the characteristics of military deployments or missions during the Cold War.

Even military systems have changed, despite the length of arms programmes usually underlined in economics and political science. When one compares today's systems, there are many features that reflect disruptions in the field of doctrine or technology. Even though the so-called 'revolution in military affairs' did not completely transform the way military operations are conducted, many changes have occurred over the past three decades in terms of systems design, military doctrine and deployed capabilities. These changes are based not only on technological trends, but also and mainly on lessons learned through conflicts and operations and systemic disruptions.

All these transformations and changes have profoundly modified the economic and social fabric of the realm of defence. The nature of armed forces' missions tends to evolve, resulting in a double movement of 'spatial unbundling' and globalization of their missions. There are also evolutions in the public/private boundaries as a consequence of policy choices, administrative reforms and the search for greater organizational efficiency. These evolutions have been widening the scope of defence-related activities and modify the organization of relations between armed forces, industry and local economies as well as the society more globally.

One traditional difficulty when dealing with defence issues results from the blurred boundaries between the realm of defence and the rest of economy and society. Indeed, things are not always clear cut, and one needs to carefully manage the way to analyse defence matters. This question appears even more difficult today for, at least, two reasons.

First, boundaries between the realm of defence and the rest of society have become more and more difficult to determine. Many parameters or criteria formerly used to distinguish defence and non-defence matters have disappeared because of the integration of both spheres, which were quite separated or segregated until the end of the Cold War. This does not mean that defence cannot be isolated from civilian matters, but this analysis requires to reconsider how to speak about defence. It is therefore necessary to have a closer, newer look at defence issues to be able to identify what the realm of defence is today.

Second, this realm has moved and changed shape. Some defence-related activities or features no longer exist when we look at bases, industry, capabilities, civil-military relations… and new dimensions have emerged within existing activities or through new ones: military forces are committed to new missions like peacekeeping, security or cybersecurity; and there is a deeper involvement of civilians in defence support, particularly because a larger and larger share of defence-related activities is outsourced since the end of the Cold War. If one looks at the realm of defence with a 1980s approach, there is an important risk to misunderstand its true features and also to miss new dimensions of defence matters. In this respect, defence has become a new terra incognita for economics and social sciences because of all transformations it has been through over the past three decades.

This is the reason why it is necessary to revisit known areas to identify changes and explore new, uncharted ones to be able to understand what defence is today and what stakes must be analysed and understood. Facing evolving boundaries requires to reconsider existing maps and draw new ones to guide our approach of defence in the twenty-first century.

This book gathers contributors from economics but also political science and geography to explore the evolving boundaries of defence issues. It provides an overview of changes that have occurred over the past three decades and the ones which are expected in the forthcoming years, especially because of decreasing military expenditures as well as the lessons learnt through recent or ongoing missions and operations.

Indeed, it is necessary to redefine what the boundaries of the realm of defence are today. Part I of this book explores dimensions characterising the realm for defence through the economic literature (Jacques Aben), but also in terms of new boundaries with the rise of public–private partnerships (Renaud Bellais) and the economic and spatial dimensions of defence support (Josselin Droff). Another key dimension is to understand how defence can contribute to economic and human security (Jacques Fontanel and Bénédicte Corvaisier-Drouart).

One important transformation of defence issues is related to its spatial footprint. This is the reason why Part II focuses on military bases. Here, two studies look at the insertion of defence activities into local economies through the examples of Taranto in Italy (Francesca Artioli) and Brest in France (Thierry Sauvin and Roland de Penanros). A broader perspective is then provided on the place of defence activities through the concept of military landscapes (Rachel Woodward).

Human dimensions of defence and security have deeply changed since the 1980s. Part III is therefore dedicated to building up the human resources of armed forces (Alfredo R. Paloyo, Cindy Williams) and the place of civilians in conflicts (Liliane Perrin-Bensahel and Steven Coissard).

Part IV focuses on the international landscape of defence. Even though defence has an international feature by essence, the internationalisation of defence matters is a key dimension of the past three decades. Among several questions, the dynamics in defence expenditure is a crucial dimension (Carlos Seiglie and Scott Yi-Chun Lin) as well as the more and more international footprint of defence firms resulting from firms' strategies (Renaud Bellais and Susan Jackson) or states' requests through offsets (Gueorgui Ianakiev).

Indeed this book does not try to cover all new features of the realm of defence, but it provides a new framework to understand today's issues and helps go beyond the traditional approach of defence in economics and social sciences.

PART I
REDEFINING THE BOUNDARIES OF DEFENCE REALMS

WHAT REALMS FOR DEFENCE ECONOMICS?

Jacques Aben

ABSTRACT

This chapter has only for an ambition to introduce the present book with obiter dicta about the bridge between defence 'realm' and economics 'realm' named 'defence economics'. Especially to non-French readers it proposes some insights into the French defence institutions and policy which could be startling, by nature and by the fact that the French reality is rarely presented in defence specialised journals. To realise the introductory function, the chapter reviews 100 papers, which were available online at the beginning of 2011 spring, and tries to classify them along the lines drawn from a tentative to define precisely the concept of defence policy, for the sake of the too rare students interested in defence studies.

INTRODUCTION

When studying defence matters, one may wonder how to delimitate this field of research, especially since it goes well beyond economics. One may then ask what the 'Realms' of Defence[1] are. It's certainly not by chance

The Evolving Boundaries of Defence: An Assessment of Recent Shifts in Defence Activities
Contributions to Conflict Management, Peace Economics and Development, Volume 23, 3–24

ISSN: 1572-8323/doi:10.1108/S1572-832320140000023001

that the editor has chosen this title when designing a project for defence specialists. These specialists had not only to belong to the scientific community of *defencics* or *defensology*, they had also to be economists, political scientists or geographers. However, as an economist, I must raise the question 'What Realms for Defence Economics?' Indeed, what operative meaning to give to 'defence', 'economics' and 'realms'. Moreover it is not the end of the story because afterwards these same editors chose to focus this book on the 'evolving boundaries of defence'. So it will be desirable, for a conclusion, to try to find at least one of those limits.

Defence

May be it is legitimate to begin with 'defence', even if anybody, anywhere, would be certain to know precisely the meaning of a word used every day if not every hour. However one thing is to know how to self-defend or to defend one's honour; another is to define defence as one part of a (state) policy. For a Frenchman at least, who, as belonging to a so-called Cartesian community, that is accustomed to try defining rigorously any concept, the definition of 'defence' seems to be a challenge.

The first time (after World War II, at least) the French Republic tried to do so was in 1959, during the period when state institutions were redesigned in the line of the new born Fifth Republic. In the first article of edict 59-147 on 7 January 1959, one reads: 'the aim of defence is to assure, in any time, in any circumstance and against all sorts of aggressions, the security and integrity of the territory, as well as the population's life …'. By this definition it was intended to assert that the posture of defence was permanent and global, mobilising all ministerial departments. All of them had to be prepared for the time of war or crisis, but four were nominally dedicated to defence. They were the ministries *des armées (ministry of defence)*, *des affaires étrangères (ministry of foreign affairs)*, *de l'intérieur (ministry of home affairs)* and *des finances (ministry of finance)*. They were seconded by four others, in charge of resources: transports, telecoms, agriculture and equipment.

Roughly half a century later, things changed a little bit through *Le livre blanc sur la défense et la sécurité nationale* promoted by President Sarkozy in 2008. In this defence and security review pretending to increase the efficiency of French defence policy, it was written: 'The aim of defence policy is to assure the integrity of the territory and the protection of the population against armed aggressions. It contributes towards fighting other threats able to challenge national security …'.

What has to be remarked in this new definition is the changing of 'all sorts of aggressions' in 'armed aggressions'. If necessary, this change is reinforced by the idea that defence policy 'contributes to' – of course with other policies – and no longer 'assures' national security. In other words 'defence' is no longer global but only military.

In fact this particular semantic problem is only a consequence of a want, conscious or unconscious, to be coherent. It was not coherent to have a definition of defence with a global spectrum and one ministry of defence only in charge of the armed forces. One more time the problem comes from the tyranny of 'political correctness'. On the pediment of the building hosting the French ministry of defence is engraved '*Ministère de la guerre*' (war ministry). This was the real name[2] from 1589 to 1932 and for several periods between 1932 and 1974. During this same length of time there were '*ministères de la défense nationale*' (ministries of national defence) and '*ministères des armées*' (ministries of armed forces); the expression '*ministère de la défense*' (ministry of defence) appeared only in 1974.

When World War II appeared a real risk and after its end, it was no longer acceptable to have a ministry of war, which could give the idea elsewhere that the present government would have an aggressive policy. *Pour la petite histoire* a late consequence of this attitude has been the transformation of the *Ecoles de guerre* of the three services in one *Collège interarmées de défense* (Defence Inter-Services College) in September 1993. Perhaps it is a sign of the times that this CID has been renamed *Ecole de guerre (War School)* in January 2011,[3] at one moment when French armies are engaged in Afghanistan and Libya, at least, for high-intensity combats.

Nevertheless linguistic precision and clarity can be collateral victims of this search of coherence. Having strictly delimited the use of *défense* the writers of 2008 French defence and security review needed another name for the ministerial services in charge of national security and especially for those depending of the ministry of home affairs. They have thus chosen the word 'security' one more time so that security is, in this case and not in the case of defence, simultaneously the objective and the means to reach it: to maintain the so-called *sécurité nationale* (national security), one depends upon security services except in the case when security is threatened by armed aggressions, in what particular case one depends upon defence.

So what? In political life ambiguity there is, ambiguity there will remain. But for the researcher and moreover for the teacher, it is useful to get out of the ambiguity and, maybe, to design a definition more in line with observed practices. The following one is proposed to reader's critics:

> Defence consists in the capacity to implement, in any time, in any circumstance, the means to prevent, deter or fight any form of aggression in order to guarantee the integrity and safety of the nation's interests.

This definition tries to reconcile both nature and function of defence by the simultaneous use of the expressions 'consists …' and 'in order to …'. It keeps the idea of permanence of the defence posture: any time, any circumstance. It specifies the three successive stages of defence action: prevention, deterrence, repelling, because there are risks, then threats and eventually aggressions to deal with successively. This definition stresses that one does not die only for a given territory and a given population, but for a lot of other interests; and if one does not necessarily die, because all interests do not justify so high a sacrifice, each of them has a value justifying nevertheless a certain level of sacrifice. 'Integrity *and* safety': one more time the simultaneous presence of both terms implies that defence has to begin very early to remain limited but in the end rejects compromises. To safeguard safety – this feeling of tranquillity experienced in the absence of danger – one can be tempted to abandon integrity, as in a Munich compromise; on the contrary one has to abandon the idea of safety – that is, if necessary, accept to fight – for guaranteeing the integrity of more fundamentals interests.[4] Eventually this definition chooses the global point of view: 'any form of aggression': territory and population, of course, but also economic performance, national culture, environment … have all to be defended with appropriate means, from a simple official protest to a massive retaliation.

This being accepted it becomes possible to derive a definition of defence policy:

> Defence policy consists in the decisions of a competent authority – state, group of states, parts of state – aiming at creating, maintaining and, if necessary using the capacity to implement ….

Maybe it is not useless to stress that '*gouverner c'est choisir*'[5] (governing consist in making choices) and that to define and implement a policy one has first to take decisions, which in turn will imply actions. Faced with adversity the nature of these decisions depends upon what is named in French *esprit de défense* (the spirit of defence) or, as Tucydides said, the 'native spirit of our citizens'.[6] It is the presence or the absence of this particular spirit which determines the choice between 'shame' and 'war' as it will be seen infra. The definition opens the policy spectrum to military alliances and to decentralised authorities, which is all the more necessary that defence policy is more global. If a local government is competent in economic regulation, it may design policies for strengthening the firms located on its territory. If a member of a military alliance is attacked the other members have to use 'all means' to help it.[7] Finally, to be ready to use the means of defence when the moment of truth comes, one has to create and maintain them without being sure this moment will ever come, and that means accepting to choose.

Defence Economics

Here we are: create and maintain the means of defence on one part, governing by choosing on the other part. If 'the economy is the science which studies human behaviour as a relationship between ends and scarce means which have alternative uses' (Robbins, 1945),[8] there is plenty of room for research in Defence economics, precisely because there is to explain why a government chooses or not to create and maintain the means of their nation's defence and why they decide, or not, to use them when the moment comes. To illustrate this point what a better quote than this one of Winston Churchill: 'Owing to the neglect of our defences and the mishandling of the German problem in the last five years, we seem to be very near the bleak choice between War and Shame. My feeling is that we shall choose Shame, and then have War thrown in a little later, on even more adverse terms than at present'.[9]

Knowing the ability of economics to pretend covering all fields, from labour market to the gross domestic product, via art, health, education and, why not, crime or love, there would be no reason why 'defence' could remain out of its reach. Moreover knowing that, according to the Stockholm International Peace Research Institute,[10] world defence expenditures, understood as military expenditures, amounted to 1,738 billion USD in 2011, that is 2.5% of GDP, this topic ought to be a plump prey for the economic community. Nevertheless, we have to admit that in many countries and particularly in France, economists have not been the most interested by this field, maybe because academics tend to consider defence as a 'dirty' thing.[11]

The 'Realms' of Defence Economics

Why then speaking about 'Defence and its realms'? Of course it is an aesthetic but ambiguous too, especially perhaps for a non-Anglo-Saxon person, for whom 'realms' would be able to keep all its meanings whatever the context. So we can naively make a tentative list of these meanings.

First of all, what are the *kingdoms* of defence economics? Here, perhaps surprisingly at first glance, this will not designate the countries where the thesaurus of this special branch of economics was constituted. These kingdoms will rather be those where the events of peace and war related in the newspaper have happened. The map drawn in this way ought to represent a sort of indicator of violence in our world, excepted for the fact that a number of papers are historical ones and other ones, being econometrical, use a long-term statistical material.

Speaking about kingdoms it is natural to ask who are the *kings* – and *queens* for that matter – who have compiled the thesaurus. By that suggestion we do not aim at realising a sociological study but rather a geographical or academic one in the sense of knowing where those queens and kings have chosen to go for working in that special field. It is not a pure gratuitous question, because we know that one can find countries or at least universities where there is some reluctance to study such a subject.

Secondly we know approximately what the field of defence economics is, but it remains to know what those who have ploughed this field have made of it. In other terms: what are the *topics* studied by those who have founded the economics of defence? It is not a question here to make an exhaustive and detailed review of works able to be classified in this domain, all the more so because one can find several surveys of the economic literature (Coulomb, 2004; Hartley, 2010; Hartley & Sandler, 2007). Nevertheless it may be pertinent to identify what is up to date in the field.

METHOD

To attempt answer the questions aroused during the precedent discussion, under the conditions evoked, it seems efficient to find an intermediary source. The one which seems to impose itself to the onlooker is, of course, the journal *Defence and Peace Economics*, entirely devoted to this field since 1990, and as far as we know the only one in such approach. Of course the journal's contents have significantly changed over those twenty years and if it is a question of what is up to date now, it would be a mistake and a waste of time to make an exhaustive census. Consequently, the choice has been made to take the last hundred papers accessible online at the journal's 20th anniversary[12] which means going from the first issue of volume 22 to the first issue of volume 19 included.

One thing is to have the papers at one's disposal; another one is to classify them. An inductive approach is quite possible, at least ex ante. It is certainly pertinent to let the authors draw the 'great design'. In that case what was written previously about defence would be of no effect. So it will be better to build a classification drawn from those previous definitions and then only to try to put all the papers found in the different cases. One may expect that it will not be easy and not without ambiguity, because the same paper can have several targets.

As it was said, defence policy is a set of decisions destined to constitute, maintain and if necessary use the means to prevent risks, deter threats and repel aggressions. So the first item of our classification could be on the risks, threats and aggressions.

As we know, these decisions are taken through the filter of the *esprit de défense* of decision-makers; but, without ignoring this dimension, it appears probable that it cannot be one of the principal subjects treated by economists, who are not interested in complex psychology: defence actors too can be 'bulls' or 'bears'.[13]

To be able to take decisions and make them effective, one needs first an organisation, and in fact a triple one: political, administrative, operational. This means: who decides; who receives these decisions and cast them to the interested bodies; who, at the end, acts to get the expected results. Even though it should be easy to identify the right institutions and decision-makers, the reality appears more complex and quite often ambiguous, as underlined in the case of France.

Any French person interested in these questions knows that the French institutions have suffered, and may be still suffering potentially, of an ambiguity on who decides.[14] If the President is the chief of the armed forces, if he is responsible of the independence of the nation, if even he is the authority in charge of the nuclear deterrence, the Prime minister is 'responsible of the national defence' and, as the government's chief, he 'has the disposal of the armed strength'. It was especially sensitive in 1986 when President Mitterrand and Prime Minister Chirac had to go along with each other and, to a lesser extent, between 1997 and 2002 between President Chirac and Prime Minister Jospin.

Of course the question of organisation concerns not only the central state, as it is implicit in the precedent description, but also the international organisations: military alliances or security systems, and to a certain extent the decentralised authorities.

Having designed an organisation is not enough if it has not the appropriate resources. And one can understand that the interest of economists is greatly increased at this point. As it is principally question of policy, resources means first budget or public finance. Since the famous 'How to pay for the war?' (Keynes, 1940a) and, of course, long before that, numerous were the economists to study all the special financial tools developed by the defence administration so that, at least, the financial regulations do not impede the defence action and, better, especially when the urgency commands, serve it. It seems that today, in a context of economic crisis and huge public debt, these tools have been multiplied and one uses willingly the expression 'innovative financing', which is a manner to say that the red lines of financial orthodoxy can be revised, so that, for instance, private capital may be associate with public funds to buy the military equipment of the nation.

'Resources' does not mean only money, as the Spanish Kingdom has learned at its own expense during the 16th century (Bodin, 1568). The organisation previously described needs manpower for all its levels and especially for the operative one. There has been, and there is, a lot of original studies to do about this special sort of manpower who are the military people. They have a very special job and often their recruitment can go through exotic method: for instance one author has tempted to prove that an army with draft functions quite similarly as a slavery economy and another one that the reserve servicemen act like moonlighters (Mehay, 1991).

Who says manpower says also training courses and continuous training. And it is not a superfluous question. First there is a problem of choice between two recruiting ways. On one side one can recruit young people and train them on academic and operational levels while they still are 'soft wax' so that they can be cast in the official mould. On the other side it seems efficient to take older people, already qualified whatever the level, and train them only on operational ground, accepting to relativise the motto 'I want to see only one head'. There is in fact no obvious solution: French armed services have chosen the first option; it seems that it is the contrary for the British ones.

Second a military school is a political instrument, a possible instrument of influence upon foreign cadets. So in the 1960s the French and German air forces wanted to train their fighter pilots on the same aircraft, Alphajet, and in the same school. But the Pentagon objected, because it was inconceivable that German fighter pilots could be influenced by a country which had partially recessed from NATO.

The best warriors cannot do anything without good weapons, and more generally, good equipment. This is, of course, the question of goods and services bought with the defence budget, their technological level, their performances and, logically, their price: everybody knows the 'gold-plated weapons' aphorism and the 16th law of Norman Augustine.[15] And, as for manpower, there is a corollary question: produce or buy? It is more efficient to buy, especially for a small power. Russia or the United States (and some others of course) are ready to sell, more or less, at a very competitive price due to the economies of scale in arms production linked with their own huge purchases. But at what price? Who does not remember the French aircraft grounded by the lack of US spare parts during the Musketeer operation in 1956 in Egypt; and the threat to stop the US logistical support to the Israeli army during the Kippur War in 1993? Of course it can be very costly to produce one's own weapons, especially without the economies of scale provided by large production series, but it is the price of independence, as was understood by the Chairman of the French council of ministers, Guy Mollet, in 1956, after the Suez campaign and what he called 'the US betrayal'. Betrayal or not, 1956 marks the beginning of the revival of the French defence industrial base.

The build-up of a defence industrial base is linked to the question of becoming an arms exporter hoping to lengthen its production series and get the profits of the Wright law on learning-by-doing, but suffering an international disgrace.

And while trying to reduce, at any cost, the defence burden, why not to grant the defence of the realm to the large, private companies? They were, in France, medieval ancestors of today's Halliburtons and other Sandlines. Because the defence level of activity is subject to important fluctuations, possibly cyclical, it would be certainly efficient not to have a permanent defence system but to be able to revive this system at the moment of truth. It was the case when the Roman or Athenian peasant described by Smith (1776) was called by the Republic. He left his plough, took his armour and his sword and went for repel invaders. And when the battle was over, he went back to his field. It could be the same thing with the hiring of military companies, which are paid only when they fight, but are supposed to continue fighting after the battle is over, because there are other battles elsewhere.

As Jean Giraudoux, a famous French author, puts it in Cassandra's mouth: 'it was the last [war], the next one is waiting'.[16] So the large companies can hope to live and make profits and the hiring states know that the cost of maintaining the capability to fight will be shared with several other states. But it is possible, desirable certainly, that the next war keeps people waiting. And in that case, who is going to pay for the survival of these companies, without challenging all the previous reasoning?[17] Anyway, as surprising as it could appear, defence production could be outsourced.

At this point it seems possible to synthesise the exposed system through a simple figure (Fig. 1). The decisions taken are the nodal point of the system, and the link between the demand of defence or of defence expenditures, on the left, and the supply of defence, on the right. The state as the nation's tutor, knows the security needs and tries to satisfy them by expressing a demand (budget) and offering a supply, alone or through public–private partnership.

The story does not end at this point, because the state's decision to hire manpower, buy goods and services and, as a consequence, levy taxes and borrow funds cannot be without effect upon the economic system (Fig. 2). This opens new realms through the study of the performances of the

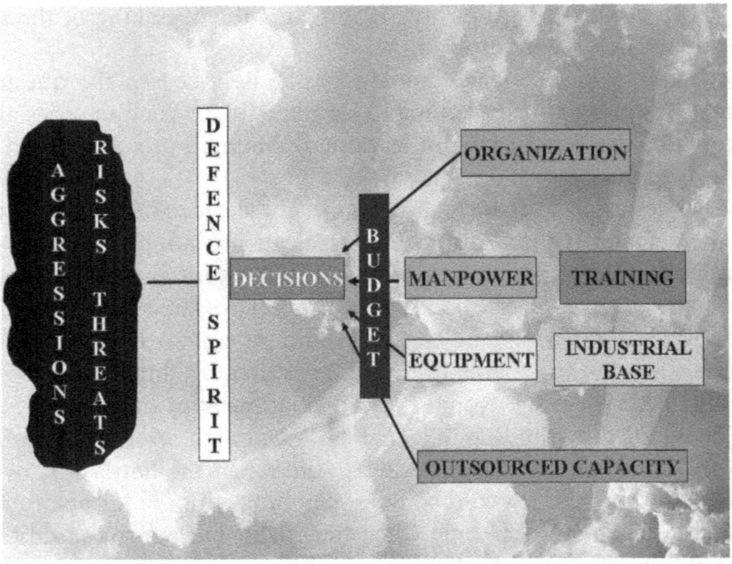

Fig. 1. Defence Policy.

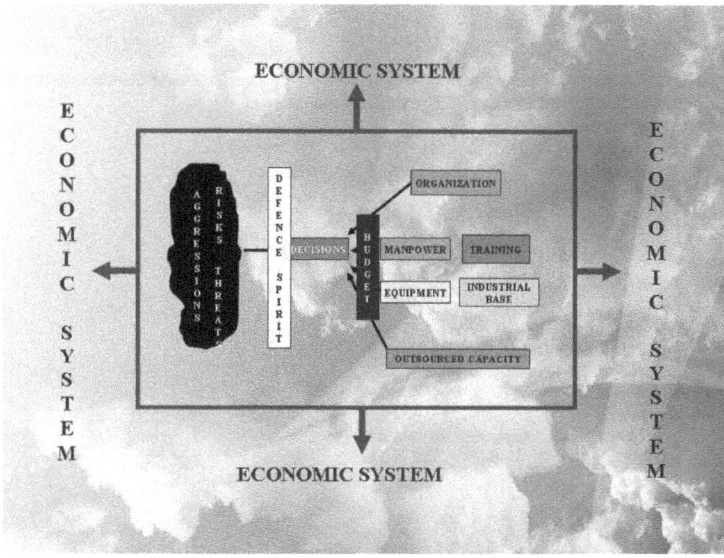

Fig. 2. Economic Spin-Offs of Defence.

economic system, as influenced by defence decisions: inflation, growth, balance of commerce, debt, (un)employment, investment … 'What else?' Precisely, it is what the survey ought to show.

RESULTS AND COMMENTS

What Kings, What Queens?

In fact Fig. 3 shows only the nationality of universities to which researchers belonged when writing papers about defence economics. It is not surprising that the United States are the most cited country. It is certainly a question of number but it is probably too a question of inclination. The interest of American authors, and universities, for defence and strategy is well known, in many countries and especially in France where regularly defence authorities underline this example to the French researchers, complaining about the under-production of such studies on this side of the Atlantic: 'Research in human and social sciences dedicated to defence and security questions

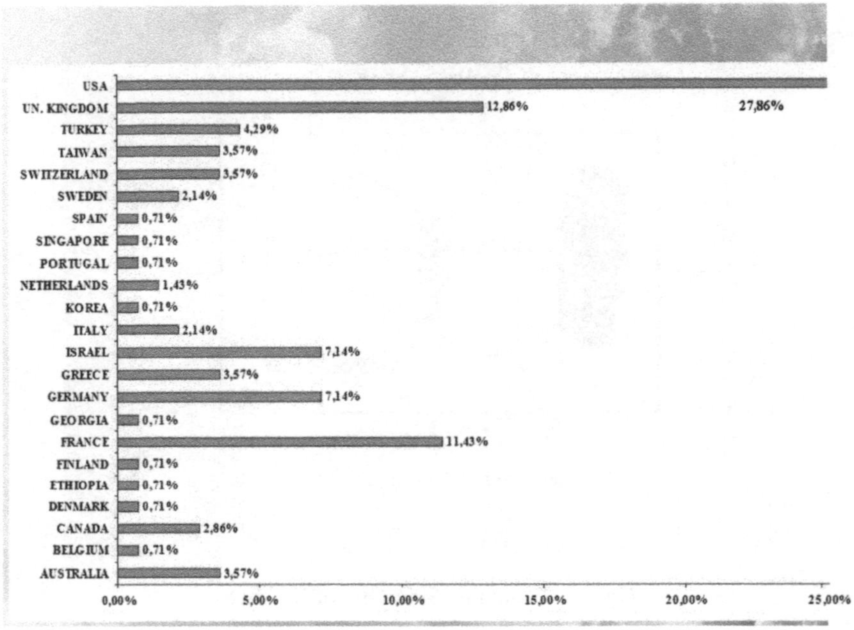

Fig. 3. Kings and Queens.

do not contribute enough to the building of a scientific culture in this realm' (Mallet, 2008, p. 307).

Nevertheless France seems well-represented in this panel, with 11.43% of authors. In fact this is a pure coincidence, because one issue of the journal was edited by professor Fontanel, of the University of Grenoble, who invited essentially French colleagues. Apart this special example, there is only one author declared as belonging to a French university.

What Kingdoms?

These 'kingdoms' (Fig. 4) are countries or group of countries cited or studied in the sample of papers. Of course this group is very heterogeneous because there are a lot of reasons to choose a particular country for a study. It can be the theatre of a war (Palestine or Iraq), an example of a special policy (USA or EU), a case of financial problem (NATO) .... Here, with Palestine, we have an eternal conflict with nothing new on the military

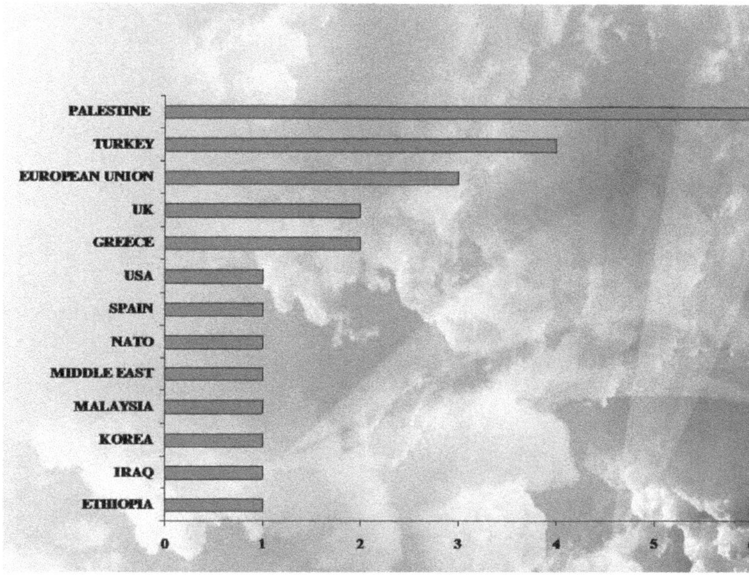

Fig. 4. Kingdoms.

ground, or else with the new interest upon terrorism. But studying the effects of this generally low-intensity conflict upon both protagonists' economies can be very profitable.

What Topics?

As we have seen previously, this third question has to be treated step by step, following the different structural components of the defence policy definition. Five topics or group of topics can be roughly identified.

Risks, Threats, Aggressions

One cannot be surprised that, among the different risks, threats and aggressions studied in our sample, terrorism takes up an eminent place (Fig. 5).[18] Nevertheless this represents a remarkable evolution since the initial issues of *Defence Economics*: at the beginning of the 1990s, terrorism was completely absent of defence economics. The large number of papers related to 'civil war' is not surprising too because of the number of those conflicts in

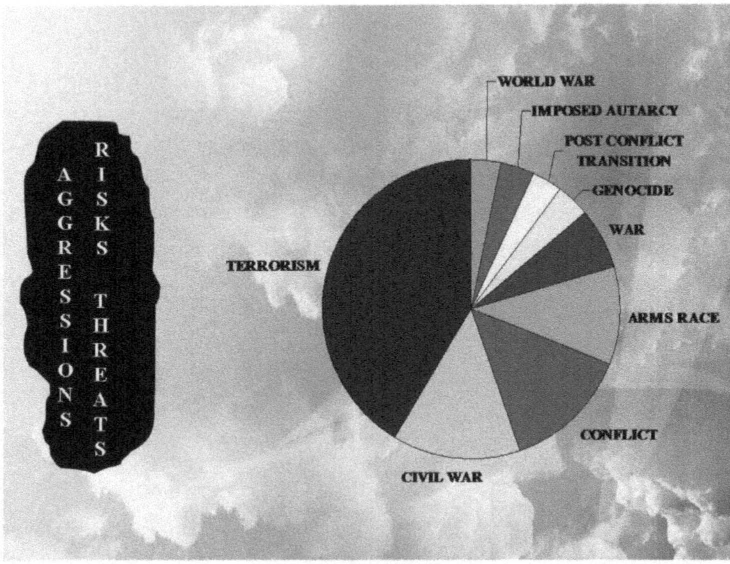

Fig. 5. Realms: Risks, Threats and Aggressions.

the world since the end of the cold war. The two items 'conflict' and 'war' are hotchpotches and do not matter here. On the opposite, 'arms race' can represent a surprise because this topic was very present in the 1980s and one may expect that everything had already been said. 'Post conflict transition', even with a small representation, can be underlined as it is a question occupying more and more the research of political scientists as much as of politicians, but seemingly this represents a field of research underdeveloped in economics.

Organisation

Because of the catch-all character of this word, and the catch-all function deliberately given to this category, the list of items ranged under this title has to be heterogeneous (Fig. 6).

So alliances keep defence economists quite busy, as it was the case during the 1970s and 1980s, when the question was to know if small nations were exploiting larger ones (Olson & Zeckhauser, 1966). Of course, new alliances or coalitions appear or are revitalised, as NATO with Kosovo, Afghanistan and recently Libya.

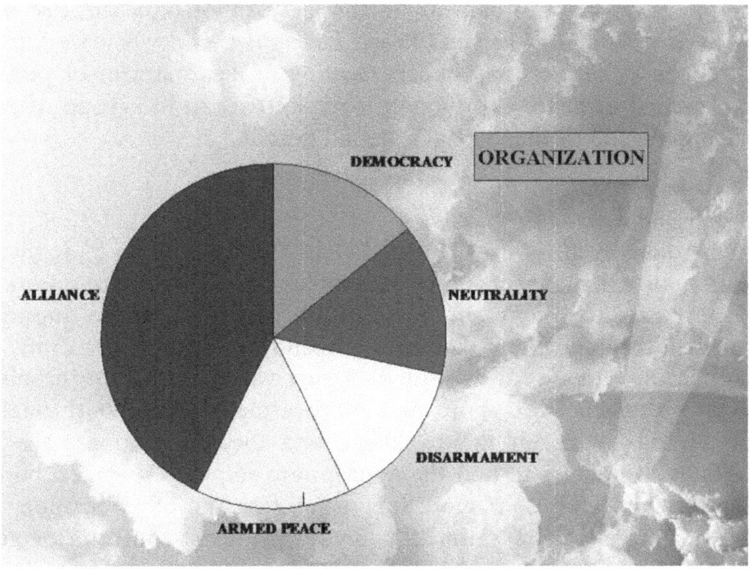

Fig. 6. Realms: Organisation.

Perhaps the new element here is the specific focus placed on the European Union as a military actor. Since 1992 and the Maastricht treaty, the EU, still a modest hard power, has been able to be involved into a number of conflicts either for military missions (Democratic Republic of Congo, Chad) or for 'rule of law' ones (Kosovo, Georgia). And defence economists are more and more interested by the military expenditures of countries which would like to exist militarily but have not the will to accept the necessary burden.

Defence economics has the same aim when looking at 'neutrality'. How this political orientation does affect the military expenditures of a country? Of course Switzerland is an ideal candidate for this sort of study, knowing the importance of military culture in the structure of the Swiss society and of military policy in the political preferences of Swiss people. It is significant that in Switzerland the approval of the last military reform, named 'Armée XXI', was submitted to a voting or referendum,[19] what is totally unthinkable in any other democracy. And, actually, one could make a very similar comment about the occurrence of 'democracy'. Does this kind of political organisation have any influence upon the demand for military expenditures?

The presence of 'disarmament' in this list cannot be a surprise with a sample drawn from *Defense and* Peace *Economics*. But when we look closely, it appears that the only papers treating of disarmament or peace are either classified in history of thought or belonging to the group of papers celebrating the end of the journal's second decade.

Manpower and Training

One more time there is no abnormality in the list in Fig. 7. It gives an impression of exhaustiveness: All the topics linked with the question of manpower seem to be present. First of all, of course, is the question of enlistment. More and more countries suspend or suppress the draft, especially when they are not at war for their survival, which is not the case for Israel for instance.[20] So, more and more armed forces go through the labour market to recruit their soldiers, and they have to use the same leverages as companies to find the appropriate employees: wages; bonuses; social programs ... and, far more surprising: freedom of association. And, having to enlarge their recruitment sources, they must confront the gender question and rule out the old bans targeting for instance women and gays.

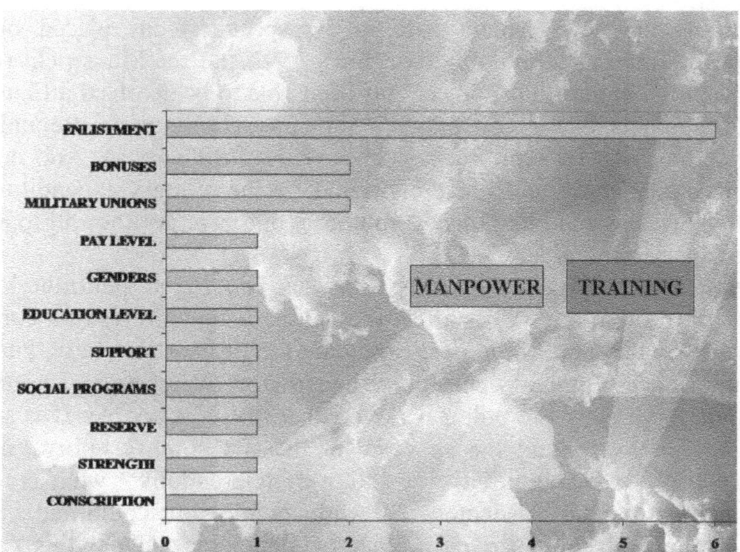

Fig. 7. Realms: Manpower and Training.

Compared with this ordinary mode of recruiting, traditional military solutions, conscription and reserve, seem to be there only as a benchmark. The only paper about military reserve analyses the effect of activation upon reservists' earnings: it appears that contrary to some expectations reservists have a financial interest to be activated regarding their average civil earnings. One may note that during the whole life of the *Defence and Peace Economics* journal, the number of papers treating of reserve issues has remained very small, but it is the same thing in the other non-economic journals.

Equipment and Industrial Base

The more or less general studies about the industrial base take the lion's share in this group (Fig. 8). There is nothing astonishing for an aggregated category, but nothing astonishing either regarding restructurings which affect the defence industry. These papers concern the United States, of course, the United Kingdom, but through a 'country survey', and other countries from the European Union. One more time this shows that European defence begins to be a serious matter for academics. However, in

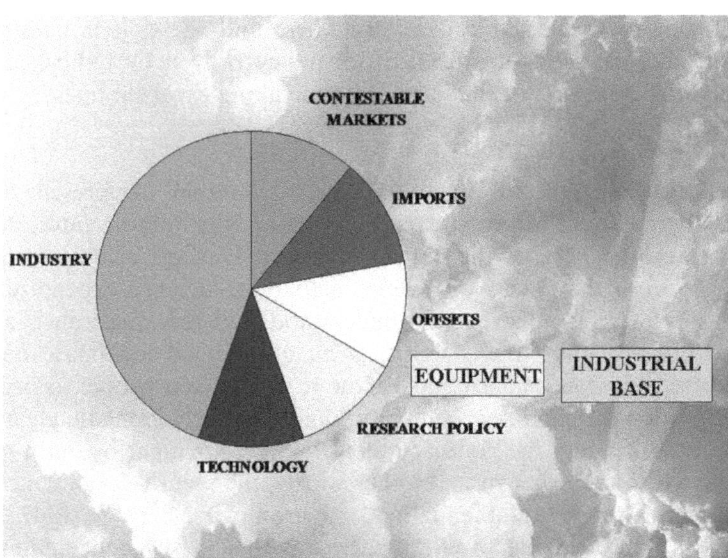

Fig. 8. Realms: Equipment and Industrial Base.

this case, the assessment is that the European defence industrial market is inefficient: not a surprise but still a pity.

'Technology' and 'research policy' are certainly closely linked but here we have only two papers, one treating the research policy of the United Kingdom and reviewing the key issues faced by the MoD and the other one studying the effects of a network organisation of the industrial base upon the innovation. In the second example, a network centred on a defence agency and principally formed of system integrators raises the control of relevant technologies but reduces the innovation variety.

In itself the 'contestable market' did not deserve a special class, especially because there are only two papers including such concept into their title and, in one of them, the term industry is used in its general meaning. But the issue is of interest as it is stated in the paper that there would be a negative correlation between the size of industrial sector and the probability of civil war. This is certainly a result that ought to be explained to Yugoslavian peoples. More seriously, the second paper tries to apply the theory of contestable markets to a case out of its perimeter – state-regulated markets – and it appears that the defence market can be, at least, quasi contestable due to an adequate state policy.

Offsets in arms transfers is a traditional issue, but it seems better dealt with by newspapers than academic journals. Indeed it is a topic where it is very difficult to find reliable data. But, here and there, it is possible to establish links between an official offset policy proven by published contracts and the evolution of the local economy along expected lines.

Economic Spin-Offs

Fig. 9 speaks for itself. All possible economic variables are present on the list. But there is a clear 'growth tropism' among the authors interested by defence spin-offs: 12% of the papers reviewed – except errors and omissions – concern the effects of defence activity or defence expenditure on economic growth. It must be well understood that this question is a very sensitive one, because it was born as an attempt to legitimate defence expenditures on economic grounds. How to convince a people to bear the military burden during a peaceful period and especially immediately after a war? Because it would be good for the GDP, employment or, on a global perspective, economic growth. In other words, the security argument, even correct, is no longer audible. It is the reason why, even in 1941, John Maynard Keynes looking for the Americans to join the fight against the Axe had to write: 'It appears to be politically impossible for a capitalistic democracy to organize expenditure on a scale to make the grand

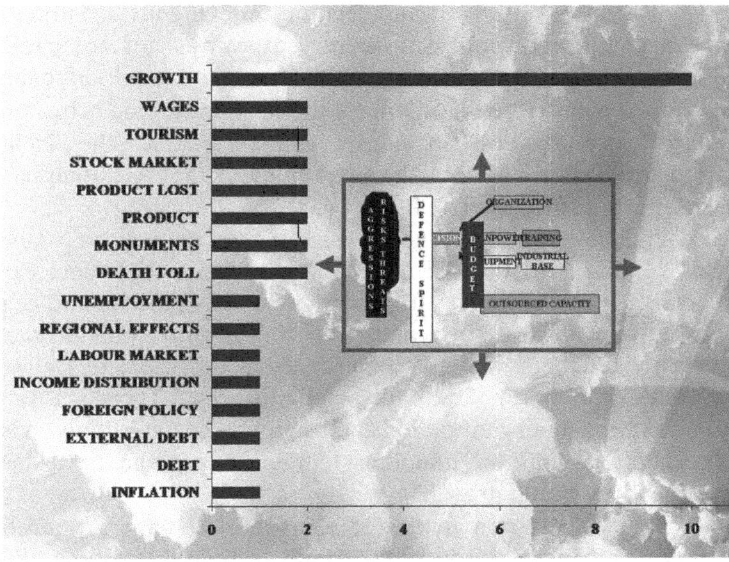

Fig. 9. Realms: Economic Spin-Offs.

experiment which would prove my case except in war conditions' (Keynes, 1940b, p. 150).

What Keynes had in mind was that public expenditures needed to finance a war would be so huge that they would have a tremendous positive effect upon the global demand for goods and services, then upon employment, then finally upon national income. In some sense the game would largely be worth the candle because the demand of households nourished by public expenditures would add to the first effect and justify a new supply of goods and services and a new raise of employment and income.

But is it possible to maintain permanently a war economy? Is it possible to choose definitively 'guns' against 'butter'? Where to find the resources to finance both defence expenditures and private investment? How to produce the goods and services demanded by households without investing for their production? And what if defence expenditures eventually depress the growth rate?

These two visions of the effects of defence expenditures are translated into the conclusions of the different papers treating of this subject.

The problem is that among 10 papers found, 5 detect a positive relationship between defence expenditures and growth, 4 find a negative one and the last 1 does not see any significant relationship. Of course, Turkey is not a European country; of course a country at war is not comparable to another one having experienced peace for a long period; of course an underdeveloped country can hardly have the same economic behaviour and conditions as an advanced economy; and of course 'the Toda and Yamamoto causality test' is not the same thing as 'a VAR analysis' or as 'parametric partial correlations'. But so what?

Already in 2005 a rapid (and unpublished) study realised on a sample of 27 articles established that 1 did not see any influence of defence expenditure on growth, 4 saw a double effect, simultaneously positive and negative, 11 concluded to a positive effect and … 11 to a negative one. So the story goes on evenly and seems able to go on for the next decades of *Defence and Peace Economics*. Who spoke about a 'plump prey'? There are so many countries, so much significant periods and so much statistical methods, that there is certainly no limit, no boundary to studies upon the 'nexus between military expenditures and growth'.

If this inquiry had been made for the first 100 papers published in *Defence (and Peace) Economics*, the results would have been different enough. And this ought to be desirable because it means that defence studies are naturally linked to the reality of defence policy and practice and that this reality is obviously changing. Of course, because defence studies are more and more a mix of disciplines, a theme which is now obsolete can remain within the academic topics, because it has been taken in charge by historians or economic historians when rejected by other scientists. And we are to be prepared to a lot of these transitions, because the debt crisis is going to reduce the defence effort, at least in Europe and may be in the United States, forcing them to find absolutely new solutions for producing more efficiently external security. It's certainly why the privatisation of defence is so *en vogue* among scientists, whatever be their discipline. In the same time, these same nations are confronted more and more to a security challenge within their frontiers, so that they have to develop their police with the spoils taken on the armies.

So without discussion, the boundaries of defence are moving and so do defence studies. What follows within the boundaries of this book is to prove it.

NOTES

1. The reader could be interested to know that the English writing has been chosen deliberately by the present author, in a sort of defensive stance against the common conformism … or as nostalgia of Victorian ages.

2. In fact *Secrétariat de la guerre* or *à la guerre* (Secretary of or for war) from 1589 to 1791.

3. Another sign is given by the fact the first promotion of the new *Ecole de guerre* is named '*Général de Gaulle*'.

4. On 6 September 1914, General Joffre while giving his orders of the day wrote: 'any troop unable to advance will have to accept death rather than to abandon the terrain taken to the enemy' (engraved on the pedestal of his equestrian statue on the Champ de Mars, in Paris).

5. Pierre Mendès France, the French Prime minister who accepted the compromise ending the Indochina war in 1954.

6. Tucydides (n.d.): 'If we turn to our military policy, there also we differ from our antagonists. We throw open our city to the world, and never by alien acts exclude foreigners from any opportunity of learning or observing, although the eyes of an enemy may occasionally profit by our liberality; trusting less in system and policy than to the native spirit of our citizens'.

7. See for instance the article 5 of the North Atlantic Treaty signed in Washington on 4 April 1949: 'The Parties agree that an armed attack against one or more of them in Europe or North America shall be considered an attack against them all …'.

8. Why this definition rather any other? This one is among the most cited and it is a normal respect presented here to somebody having written *The economic causes of war* in 1939 (Howard Fertig, New York).

9. In a letter to Lord Moyne in1938, cited by Gilbert (1991). He had the opportunity to verify this guess and say to the Commons "England has been offered a choice between war and shame. She has chosen shame, and will get war" (Keane, 2005).

10. http://www.sipri.org/databases/milex

11. At least, it was the experience of the present author with some of his colleagues.

12. There is something ironical in the acknowledgement that with the twentieth anniversary of the journal some authors have proposed surveys. But first they were not online at the moment chosen and secondly they did not, by definition, respect the condition of immediacy set down.

13. 'The market price will be fixed at the point at which the sales of the 'bears' and the purchases of the 'bulls' are balanced' (Keynes, 1936, p. 108).

14. To explain so curious a situation, one says, for instance, that in 1958 the writers of the constitution did not know if General de Gaulle would remain prime minister or would prefer to become president (the outcome being certain) after the coming into force of the constitution of the new republic. So they covered both cases.

15. 'Defence budgets grow linearly but the cost of military aircraft grows exponentially'. *Corollary*: By 2054, the entire U.S. defence budget will purchase one aircraft. It will be shared by the Air Force and the Navy 3 1/2 days each week, except in leap years, when it will be made available to the Marines for the extra day – http://philebersole.wordpress.com/2011/03/20/augustines-laws/

16. Giraudoux (1935).

17. 'On 16 April 2004 Sandline International announced the closure of the company's operations. The general lack of support for Private Military Companies (…) is the reason of this decision', http://sandline.com (8 September 2011).

18. Especially if one knows that these specific lines were written, by coincidence, on 11 September 2011.

19. On the 18 May 2003, 1.7 million Swiss people (among 4.5 electors) voted in favour of a reform reducing drastically the volume of the Swiss army and giving to it a more professional character. Only 0.5 million refused the reform.

20. Even if there would be a lot to say about the responsibility of this country on its own insecurity.

REFERENCES

Bodin, J. (1568). *Réponse au paradoxe de M. de Malestroict touchant l'enchérissement de toutes choses, et le moyen d'y remédier.*

Coulomb, F. (2004). *Economic theories of peace and war.* London: Routledge.

Gilbert, M. (1991). *Churchill: A life* (p. 595). New York, NY: Henry Holt & Company.

Giraudoux, J. (1935). *La guerre de Troie n'aura pas lieu* (p. 7). Retrieved from http://beq.ebooksgratuits.com

Hartley, K. (2010). *The economics of defence policy: A new perspective.* London: Routledge.

Hartley, K., & Sandler, T. dir. (2007). *Handbook of defence economics.* New York, NY: Elsevier.

Keane, M. (2005). *Dictionary of modern strategy and tactics* (p. 15). Annapolis, MD: Naval Institute Press.

Keynes, J. M. (1936). *The general theory of employment, interest and money.* London: Palgrave.

Keynes, J. M. (1940a). *How to pay for the war: A radical plan for the chancellor of the exchequer.* London: Harcourt Brace.

Keynes, J. M. (1940b). The United States and the Keynes plan. *The new republic*, July 29, 1940. In D. Moggridge (Ed.). *The collected writings of John Maynard Keynes* (Vol. 22). Cambridge: Macmillan. Published in 1978.

Mallet, J.-C. (2008). *Livre blanc sur la Défense et la Sécurité nationale.* Paris: Odile Jacob.

Mehay, S. (1991). Reserve participation versus moonlighting: Are they the same? *Defence Economics*, *2*(4), 325–338.

Olson, M., & Zeckhauser, R. (1966). An economic theory of alliances. *Review of Economics and Statistics*, *48*(3), 268–279.

Robbins, L. (1945). *An essay on the nature and significance of economic science* (p. 31). London: Macmillan.

Smith, A. (1776). In E. Cannan (Ed.), *An inquiry into the nature and causes of the wealth of nations* (Book V, chapter I, sections 1 to 9). Retrieved from http://www.econlib.org

Tucydides. (n.d.). In R. Crawley (Trans.), *The history of the Peloponnesian War* (Book 2, chapter 6). Retrieved from http://classics.mit.edu/Thucydides/pelopwar.2.second.html

PUBLIC-PRIVATE PARTNERSHIPS AND THE TRANSFORMATION OF DEFENCE INVESTMENT

Renaud Bellais

ABSTRACT

Launched in the early 1990s in the United Kingdom, Public-Private Partnerships (PPPs) induced radical changes in both the public-private boundaries and the production of state-provided services. Such 'budgetary revolution' impacted the biggest state spender in capital expenditures, that is, the Ministry of Defence. Today many MoDs are expected to leverage on the British experience and develop their own approach of PPPs to overcome both the ineffectiveness of their defence spending and today's stalemate in public budgets. This chapter leverages on British experiences over the past two decades to analyse the benefits and limits of PPPs in the realm of defence. Does such contractual arrangement fit defence-related investment? This chapter explores the on-going redefinition of public and private realms in military matters and it puts into relief the key dimensions of PPPs in terms of contractual arrangement.

The Evolving Boundaries of Defence: An Assessment of Recent Shifts in Defence Activities
Contributions to Conflict Management, Peace Economics and Development, Volume 23, 25–49

ISSN: 1572-8323/doi:10.1108/S1572-832320140000023005

Since the end of the Cold War, military budgets have decreased and many countries cannot afford all investment projects they would like or expect, even the United States despite a strong increase of their budget in the 2000s. European countries in particular have halved the share of GDP they dedicate to defence between the 1980s and 2010s. Nevertheless international relations are still marked by instability and conflicts, and major powers are expected to avoid major crises or humanitarian disasters.

In this context, armed forces face a dilemma. With a decreasing purchasing power, they cannot invest in all equipment and infrastructure projects they need. This is the reason why ministries of defence are looking for alternative modes of investment. This quest of innovative approaches has resulted in a reshaping of the boundary between the state and the private sector. Defence is not excluded from such an evolution that impacts the realm of defence. Indeed, the stronger involvement of firms into defence matters questions the traditional approach in terms of public goods.

Following the neoconservative revolution in the 1980s, the United Kingdom was the first country to introduce public-private partnerships (PPPs) as a means to complement public investment with private funding in defence. Initiated in the early 1990s under the concept of 'private finance initiative (PFI)', this kind of privately managed public investment has quickly become a major financial resource for the ministry of defence. Progressively, other countries have been experimenting this alternative approach to publicly funded investment.

Nevertheless as soon as they were implemented, PPPs attracted several critics. Indeed, using private funds to deliver public goods could lead to a privatisation of state missions. One could also wonder whether a privately managed service is as effective as a public one, and whether this approach can provide the expected savings. This evolution of the public-private boundary can be considered as a risk regarding the core missions of the state. It is therefore useful to question the very meaning of PPPs and their use in defence matters.

PPPs, A REVOLUTION IN DEFENCE PROCUREMENT

PPPs introduce a radical change in public investment. While private funds were historically banned from public investment, the United Kingdom reverted this principle from 1992 by favouring private investment in every field of public services and limited public investment when such a choice was not possible. This revolution challenges the traditional approach of public goods and leads to wonder whether PPPs can deliver the expected outcomes.

Redrawing the Public-Private Boundary in Public Goods

From a theoretical point of view, almost since Adam Smith's *Inquiry on the Wealth of Nations* in 1776, defence belongs to the category of public goods and often provides the perfect example of public provision for this category of goods. As admitted by Samuelson (1954), defence is a public, or social, good 'which all enjoy in common in the sense that each individual's consumption of such a good leads to no subtraction from any other individual's consumption of that good'. This definition implies that a pure public good must exhibit two characteristics: non-excludability and non-rivalry. Defence has been used as the best example of public goods for ages in economic textbooks (Table 1).

Even though this approach appears still valid, in reality most of public goods are not pure public goods. For instance, if one agent does not contribute to the financing of a given public good, it might exist a technical solution to prevent him from benefiting by this public good. In defence we can consider an imperfect geographic covering as a source of impurity of a public good, since not all economic agents benefit from the same level of utility. The American national ballistic missile defence programme represents a good illustration of such a case, for it does not expect to provide an equal protection in whatever place within the American territory even though this system is funded by the Department of Defense.

Beyond this limiting feature against impure public goods, it appears that we must consider defence not as the sum of specific systems or services, but as a global function resulting from a systemic interaction between, and beyond, all its contributing inputs. From an economic perspective defence as a whole is not excludable and thus one cannot expect that a private

Table 1. Samuelsonian Typology of Goods.

	Exclusive	Non-Exclusive
Rival	Private goods	Impure public goods
Non-rival	Impure public goods	Pure public goods

entrepreneur is bound to supply it. That is the reason why it is useful to consider defence as a pure public good.

An impressive literature has discussed different problems linked to public goods. First, the provision of public goods raises one issue: identifying what is the optimal quantity of good to offer. In the case of national security, we observe the impossibility for the State (under the condition that the size of State is large) to know exactly how much each citizen is willing to pay to have the level of public good that maximises his utility function.

Second, what are the institutions (and policies) able to supply this public good at the level of efficiency? As an entire community shares the benefit of defence, there is a window of opportunity to free ride. The theory of club goods (Cornes & Sandler, 1996) defines the optimal quantity of public goods according to consumers and the price each consumer pays. Nevertheless, policymakers have to select, through legislature, the best mechanism of providing defence, which is the closest to a Pareto-optimal choice.[1]

Finally, the provision of public good causes as many problems in the aggregation technologies of public supply as in the principles of financing. The production of defence consists of determining the most efficient means of delivering equipment and services that contribute to the defence policy. Indeed the economic literature traditionally considers that once a good is defined as a public good, the state has to take charge of it. Nevertheless this assumption can be challenged.

One must analyse separately two main functions inherent to defence (and to public goods more globally): its production and its provision. Provision of public goods is the core function of the government: warranting the security of all the people and properties throughout a given territory. This function justifies the reference to the concept of public good per se. On the opposite, production does not imply that the State can be the only relevant economic agent to deliver a given public good. Here, there is no relevant theoretical argument supporting the idea that the state must take charge of the production of defence. Such a distinction leads to wonder whether the state has to directly manage all activities related to defence matters. One may question whether the introduction of market mechanisms in the realm of defence, which was previously considered as intrinsically the realm of the state, is possible and can improve effectiveness of producing this public good.

If defence as a social function represents an example of pure public good, the components of the defence system do not share the same features. Indeed a defence system (combat aircraft, armoured vehicle, frigate ...) has a limited area of action; and it can be manageable to exclude some people from the protection it provides.

Nevertheless, defence must be considered as global function: its social function results from an integrated service provided through the whole activity of armed forces, not by an isolated part/component of them. In other words we must distinguish a social function (defence) and the means to carry it on (armament, systems and services). The distinction between a social function and the means to fulfil it has a strong consequence: it is no longer compulsory that the state takes charge of all activities required to provide defence. Its 'core competency' consists of providing defence, but it can rely on private partners to reach this aim. That distinction does not exclude any cooperation between the state and its industrial partner. The true question is to find where the frontier might be when considering a specific need.

Two main methods are traditionally suggested: public production or contracting-out (through a competitive process in public markets). The arbitrage between both depends on the relative effectiveness of each economic agent. The solution of outsourcing is largely admitted in arms procurement as the most efficient means to allocate production factors. This involvement of the private industry is not limited to armament. Many experiences in the United Kingdom, Germany and the United States have validated the idea of outsourcing a large share of military activities through various types of contracts. One may then wonder what the limit of market-based solutions is in the realm of defence and whether such solutions truly solve armed forces' needs.

In this process, the use of PPPs in defence represents an attempt to redefine the boundaries between the Ministry of Defence and the market to deliver the best value for taxpayers' money. In that sense, this new approach of public investment initiated in the early 1990s in the United Kingdom constitutes an empirical experiment that tries to test how to reshape the respective roles of the state and the market in many public services.

The Emergence of PPPs

The idea of involving the private sector in defence is not recent. For instance, the United Kingdom had yet experienced such partnerships during the 18th century when private entrepreneurs ensured logistical or supply tasks for military campaigns of the Duke of Marlborough. More recently, the defence industry has been fully associated to the maintenance for Air Force's fleets in many countries. This relationship between the ministry of defence and private firms is particularly strong in the United Kingdom.

Unlike privatisation, contracting-out and outsourcing are often used as a mechanism to reform the provision of public sector services. Contracting-out and outsourcing consists of opening up to competition a set of economic activities which were previously immune from it. Such a transformation does not necessarily entail a transfer of the ownership of assets from public to private entities.

A PPP constitutes a specific form of contracting-out. Here, the state buys services provided by a private company through an asset, which is privately owned and managed. This kind of outsourcing can be defined as a collaborative arrangement between the state and one (or more) private parties. This partnership takes the form of a contract, which specifies reciprocal rights and responsibilities, and implies some sharing of risks, costs and/or assets. The private sector funds and sets up the required assets, which are made available for the public sector. One can also consider as an elaborated kind of leasing since the state has to purchase a stream of services over a given period rather than invest in a capital asset that provides required services.

Therefore PPPs entail a modification of the relationship between public finance and public provision. Indeed, we can observe some privately provided public goods, some private goods or services bought by the Ministry of Defence (MoD). PPPs are used to exit from some activities that were previously provided by armed forces (Allen, 2001).

The creation and development of PPPs results from the fact that the state can no longer support the level of public services considered as socially desirable.[2] Through PFI contracts and other PPPs the British government and eventually other governments expect to give up their role as the driving force behind many public service initiatives; on the opposite, they look for leveraging on the expertise of private sector.[3]

Indeed, since the 1980s, many states want to take advantage of funds available to the private sector and of management practices of this latter. Even though private firms cannot get the best interest rate when issuing bonds compared to the state, they can introduce innovations that generate productivity gains. Extra financing cost needs to be offset by savings in other aspects of the privately managed project. Moreover firms can elaborate innovative financial schemes, which public entities are not allowed to use.

The British strategy was quite extreme about such a change of investment approach as soon as the government launched the PFI. In November 1994 Kenneth Clarke, then Chancellor, made clear that 'he wanted to maximise the scope and use of private finance, reserving public capital provision for those areas where private finance was considered inappropriate or could not be expected to provide value for money' (Allen, 2001, p. 15).

Among these PPPs, the Private Finance Initiative has played a precursor role since the early 1990s. PPPs were expected to represent an increasing part of public investment, up to 20% of all capital projects. As of March 2012, there were 717 current projects, of which 648 were operational, and their total capital costs reached £54.7 billion. In fact, the Ministry of Defence does not have the largest number of PPP projects: it ranks sixth with 46 projects, preceded by the Departments of Education (166 projects), Health (118), Transport (62), Communities and Local Government (64) and the Scottish Government (85). However its projects appear more capital-intensive and complex than those of other departments. It ranks second with total capital costs reaching £9.1 billion.

Therefore defence-related PPPs constitute a major commitment for the British state and, due to the size of such projects, one may wonder what the level of related risks is. The graph next page shows (Fig. 1) that from the 1990s to the 2000s, defence PPPs have grown in size and targeted more and more complex projects. While initial projects focused on simple activities (sewerage management, training facilities, family quarters, simulators ...), the most recent projects deal with advanced systems like the Future Strategic Tanker Aircraft or the Skynet 5 satellite-based military communications.

The British experience is by far the largest and most of the lessons learnt result from it. If the United Kingdom is the country that implemented PPPs on the largest scale, there are many countries that use similar contractual forms to facilitate the financing of military infrastructures or access some kinds of defence-related services. The United States has been using PPPs for decades, but in less formalised approach and an ad hoc basis, mainly to manage defence bases or infrastructures. Some major countries like France or Germany began in the 2000s to use PPPs, but in a limited approach. For instance, from 2004 to 2013, there were only 39 PPP projects signed at the national level in France and, as shown in Table 2, the Ministry of Defence only launched six of them. Nevertheless, at the international level, the use of PPPs is an emerging trend in the realm of defence because ministries of defence have to deal with budget cuts or insufficient resources.

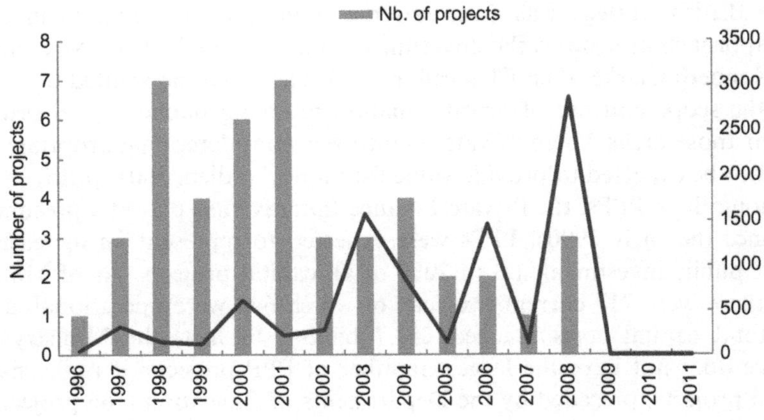

Fig. 1. PPPs Signed by the British Ministry of Defence. *Source*: HM Treasury (2012b).

Table 2. PPPs Signed by the French Ministry of Defence.

Entity	Signature	Activity
Helicopter pilot training at EA ALAT in Dax	February 2008	Training
New headquarters in Paris ('Balard' project)	February 2011	Building
IP network for Air Force bases (RDIP)	April 2011	Information technology
Caserne Roc-Noir	June 2011	Waste management and energy
CNDS Fontainebleau (sports facilities)	January 2012	Building
ISAE engineer school	November 2012	Building

Source: French Ministry of Finance, May 2013.

PPPs as a Response to Public Failure or a Financing Strategy?

If there are conditions for which the state takes charge different public services (like public utilities), they do not constitute reasons that can justify a public monopoly in designing, building and operating assets to deliver those services. The State only needs to make sure that these services are provided to taxpayers at the best 'value for money'. Pollitt (2000, p. 16) summarises this philosophy when he explains: 'The government should always investigate the possibility that the private sector may be able to provide it with cheaper and/or better quality services by increased private sector involvement in the assets which produce those services'.

Public-choice analysis explains the public failure in defence decisions by two dimensions. Government, as elected representatives, does not consider allocative efficiency and prefers to favour producers. Moreover, defence sector is characterised by a budget-maximising bureaucracy that disturbs efficient process. But the main reason of an inefficient public provision rests on the nature of contracts, the behaviour of contractors, the incompleteness of contracts and information asymmetries:

- drift of costs in traditional defence acquisition (e.g. cost-plus contracts);
- contracts without competition, favouring many asymmetries (i.e. excessive profits for the supplier, non-economic contracts);
- transaction costs for 'over-regulated' contracts versus transaction costs for uncertain contracts (entailing costs of renegotiating, re-enforcing ...).

PPPs provide an answer to improve the public provision of asset-based services. This is especially true when there exists a competitive market and a possible benchmark for armed forces between an in-house solution and an external one, for instance in car maintenance, health services, property management, support services or training centres.

What are the main reasons for a shift towards more private financing and private provision of services contributing to the production of public goods? Chang et al. (1999) underline that by combining government expertise, assets and resources with additional contributions from the private sector, PPPs can offer a wide spectrum of benefits:

- leverage assets, reduce capital investments, reduce costs, keep budgets under control to achieve projects or financial arrangements goals;
- increase the value of real estate or other assets through innovative and additional uses;
- create new capabilities and new financial opportunities that help armed forces focus on their core missions;
- influence technology early and thereby get equipment fielded earlier and/or possibly at lower cost;
- reveal the true costs of defence inputs through market mechanisms and particularly revealed costs (both inside defence and from potential competing suppliers).

According to Williamson (1985), in some circumstances, transaction costs can be large enough to bypass benefits. In this case, in-house or internal production is optimal. In the case of defence, PPPs present a substantial advantage because the private contractor includes transaction costs in the contract, what implies a shift in costs payment and leads to (un)sunk costs for the state. Nevertheless one can acknowledge that the size of transaction costs varies according to the degree of contractual incompleteness and the ownership of physical assets required for service delivery (Domberger & Jensen, 1997).

Moreover PPPs imply that defence provision evolves from the technical and operational specification of a given system or infrastructure to a service-based approach, redesigning the respective tasks of both state and private sector. In a PPP, the private operator designs, builds, finances and operates facilities based on 'output' specifications defined by public decision-makers. Such projects need to achieve a genuine transfer of risk to the private contractor to secure value for money in the use of public resources.

Providing defence is not an easy task, since armed forces must guarantee that they can carry on their missions and duties in a permanent posture. Then military customers must ensure that the selected contractor has the capability to perform the desired services throughout its contractual commitment.

Middleton (1998, p. 89) notes that contracting becomes much more complicated when a free market does not exist for certain services and when it is uneasy to evaluate the accomplishment of provided services, illustrating this idea through the case of jails: social benefits lie in having a facility available *for* use, not necessarily *in* use. The situation is quite similar regarding defence: some related services do not have a positive output, and it is difficult to define contractual performance indicators. Such a point of view can help to draw the boundary of 'PFIsable' defence activities. There is a commercial market for housing, training, logistics … but not for 'core military activities' that truly characterise armed forces and their missions.

That is the reason why the success of PPPs requires defining what kind of defence missions the government cannot (fully or partially) fulfil efficiently without a private-sector involvement. The classical taxonomy of pure/impure public goods can surely help define the modalities of financing and associated limits to reorganise defence around its 'core business'. Before analysing in a normative perspective what the state should or could contract out, outsource or 'PFIise', one must examine in a positive perspective how PPPs can meet defence needs (François, 2000).

NO SILVER BULLET: LIMITS OF PPPS IN DEFENCE

In the 1990s and early 2000s, PPP appeared as a 'silver bullet' to resolve the contradictions of public budgets, torn between limited resources and unfulfilled social needs. While privately financed publicly provided services were considered as an exception rather than an option, public institutions are increasingly required to examine potential private initiatives before engaging public funds.

Defence issues do not escape from these Caudine forks. One may wonder if such a strategy truly corresponds to the best available option. This question is especially crucial when we analyse defence needs, since these latter differ from civilian public services. Then it appears necessary to redefine the concept of PPP.

Alternative Financing or Book Cooking?

Many observers feared a substitution effect between private funding and public resources, following a deep disengagement from former public utilities and services. Nevertheless successive British governments claimed that PFI projects must not substitute to classical projects, but follow the additionality principle: PFI/PPP projects would not displace public sector funding for 'PFIsed' activities but represent additional funding that complements public one.

Actually PFI and other contractual arrangements aim at overcoming a lack of public resources and providing services related to State missions. When the British MoD started to implement this new approach, the PFI was an answer for challenges in paying for the modernisation of armed forces. Indeed most of NATO armed forces have seen their budgets reduced by a third in real terms since the 1980s peak. The last decade of the 20th century might be considered as a 'procurement holiday' and existing equipment must be replaced while budget resources are scarce.

This is the reason why PPPs appeared as a solution to solve this budget dilemma, accentuated in the 1980s by Maastricht criteria[4] within the Eurozone and by the golden and sustainable investment rules[5] in the United Kingdom. PPPs mean that more capital projects are undertaken for a given level of public expenditures, and the required investment is launched earlier than public finance would have allowed. According to Tim Stone of KPMG, the accountancy firm, 'Britain is "ten years ahead of everyone else" in the delicate art of matching private money to public needs'.[6] Its experience shows however that there is many a slip 'twixt cup and lip'. The much-vaunted qualities of PPPs as they have been developed may lead to major drawbacks.

Some PPP capital expenditures clearly substitute public investment when some public capital spending is replaced. Investment through PPPs may be additional when any second round effects are taken into account. For instance it is additional when PPP projects create new capacities or fulfil an equipment gap. Moreover when PPPs generate efficiency savings, the global budget can be used in more projects than initially expected. The real question is how PPPs are going to be implemented, and if these latter do not constitute a way to privatise public activities.

Second savings resulting from the PFI may be illusory. Indeed, the short-term cost is dramatically reduced, since the MoD does not have to invest in assets. Each PFI contract implies a small amount of money every year, since its costs are divided over decades while classical, patrimonial acquisition consists of a one-shot, big procurement expenditure. But what are its long-term impacts on armed services' ability to achieve their missions? Actually PPPs as they are currently implemented may reduce the budgetary margin of the State.

Each payment cumulates with others and the total amount dedicated to PFI contracts will eventually represent a large share of the MoD budget. PFI commitments might mortgage future budgets ... PPP projects commit the State to a continuously growing stream of revenue payments to private contractors: 'One result of undertaking long-term commitments may be a lack of flexibility in the longer-term provision of services' (Robinson, Hawksworth, Broadbent, Laughlin, & Haslam, 2000, p. 43). This can become very perilous when public administrations have to face new missions or/and realise that yesterday's choices no longer correspond to today's needs. In the case of defence mortgaging future budgets represents a true concern.

One of the biggest challenges regarding defence PPPs results from the high level of (strategic) uncertainty armed forces have to deal with. They must guarantee the country's external security and prepare themselves to unexpected threats. As experienced in recent years, geopolitical context may change quite rapidly and the defence architecture may eventually be reconceived to reply to new or evolving threats. It is therefore important that the MoD has the capacity to adapt its missions and reorganise its procurement choices.

PFI-like contracts might reduce the flexibility in investment choices because of their own features by mortgaging tomorrow's revenues. Such a situation might be worrying since short-term savings squeeze the long-term commitment to security. Signing long-term contracts with strict contractual terms may reduce armed services' margin to reorganise themselves with existing infrastructures. Even though contracts are not fully irreversible, the government 'will find it expensive to break PFI contracts if they subsequently prove to not be meeting social needs' (Pollitt, 2000, p. 19).

Third a fundamental reason why governments prefer PPPs, and especially the PFI, is that in privately financed projects capital expenditures do not normally score as public expenditures and eventually appear as 'off-balance sheet' financing. This is an interesting position, since it helps governments reduce their debt level when being able to provide some crucial needs. It seems to promote capital investment while ensuring the sustainability of public finance in the long term.

However off-balance sheet represents an 'accounting illusion', as taxpayers still have to meet the bill in the end. An honest treatment of PPPs requires taking account of such contracts to evaluate public debt. This is the reason why the British Accounting Standards Board stated in September 1998 that the capital value of PFI schemes should appear on the Government's balance sheet. Nevertheless the British Treasury negotiated in June 1999 with the ASB and issued a note that allows most PFI transactions to be excluded from State borrowing figures on the grounds that they are 'operating leases', not 'finance leases'. One may wonder if it is not exactly how Enron, the infamously famous American firm, used to manage its creative bookkeeping!

When a project is publicly funded, its costs are counted as public spending as soon as they occur. If it is privately financed, they are added to public spending years later. Such a different treatment engenders an obvious temptation ... There are some clues that governments use the PFI and PPPs to disguise the underlying position of public finances and flatter their success in managing public expenditures. As *The Economist* commented: 'To "prove" that the government is not using the PFI as an accounting scam, the Treasury constantly stresses that PFI projects involve genuine transfers of risk to private investors [...] This seems hard to swallow. Moreover, it is impossible to assess the financial impact of any risk transfer because contracts between the government and its suppliers are usually kept secret to protect commercial confidentiality'.[7]

Fourth one may wonder if PPPs really reduce taxpayers' burden and improve the economic efficiency in the provision of collective goods. On second thoughts such engagements may have the same consequences as deficit-based financing.

The Ricardian equivalence theorem[8] assumes that deficit finance has exactly the same economic impact as current taxation. Governments as well as taxpayers might then take into account tomorrow's payments corresponding to today's use of new, privately funded equipment. We could make such an analogy between State investment and PFI-like contracting. In the patrimonial approach, public investment is based on the wealth created by our parents while the PPP strategy mortgages our children's future revenues. There is certainly a crowding-out between today's gain and tomorrow's burden induced by PPPs, especially in the PFI approach.

Thus simply transferring capital expenditures between the state and private sector has no effect on the efficiency of resource allocation if identical resources are used. 'PPPs can be a mechanism for inter-generational welfare shifting', underline Parker and Hartley (2001, p. 5). If there is no efficiency improvement through PPPs in the satisfaction of social requirements, these latter appear as an accountant's manipulation to reduce artificially public expenditures. This leads to wonder what the true motivations of promoting such partnerships are.

Specific Features in the Provision of Defence

Several general ideas emerge from a review of current initiatives and experiences of PFI in the field of defence. Systematically trying to assess projects through PFI can be irrelevant, since all needs cannot be fulfilled through private supply. There is quite a dogmatic approach of the PFI, based on the conviction that such contracting is intrinsically likely to improve financial and operational efficiency in the provision of public services. While such an idea can be partially contradicted in civilian public services,[9] this is especially perceptible in defence issues because of the specific nature of the provided services.

Through the concept of value-for-money competitive forces, it is argued, will exert a downward pressure on contract prices and transfer most of risk to private sector (Shaoul, 2002, p. 3). However such an evolution cannot take place for all activities characterising public services, particularly when the logic of their implementation does not fit – partially or completely – market mechanisms. Armed forces must consider which activities are truly appropriate for private-sector market forces and profit motives, and which might be 'inherently governmental'. This raises three questions. Can armed forces always specify their requirements? Do market conditions in defence issues lead to a balanced relation between States and their providers? Can PFI-like contracts manage the defence-specific imperatives?

Under the partnership approach, project management becomes the responsibility of a private contractor charged with delivering an asset or, more and more, a global output (including systems and related services) at a fixed price. This feature does not differ from defence procurement in recent years. Cost-plus contracts have been decreasing for three decades and replaced by fixed-price contracts. Moreover there is less and less production realised in-house or by arsenals.

The true innovation of PFI/PPP in defence is the transition from product-based to a service-orientated acquisition. Many activities are transformed in a 'black box' and armed forces are no longer interested in knowing or supervising their internal mechanisms. Such a change fits quite well the current shift in procurement strategies, which have been evolving from a system designing to a capacity-building approach. This latter can help armed forces adapt to unanticipated threats or strategic disruptions.[10]

How can PPP provide an answer to such a requirement? There is a crucial issue in defining precisely the targeted output and eventually being able to make it contractible. PPPs work only when the output can be clearly described, then specified in a contract, and measured accurately. Indeed public purchasers must demonstrate a clear purpose and a strong vision of the desired outcomes from the scheme,[11] then translate this vision into a simple output specification and resist the temptation to make regular changes of such a specification.

Whoever has studied a defence programme may recognise this is a difficult point! Even though armed forces can provide a long-term, strategic view of their requirements and allow for flexibility in contracts to meet changing needs, there is a huge part of uncertainty they have to deal with. Then how can armed forces be sure that they get best value for money?

Surely PPPs can manage the basic needs of armed forces. This kind of contracts appears quite successful when treating housing, logistics, base operations or any other non-specific activities like 'white fleet' transport (Pint, Bondanella, Cave, Hart, & Keyser, 2001). Here, output can be clearly specified and a private contractor can manage to provide it more efficiently than a public administration. It is not the case for defence-specific needs, especially systems or activities located at the heart of armed forces' missions: their 'core business'. For instance, an arms programme takes at least ten or fifteen years from the initial concept to the in-service system, a time lag when technologies change dramatically and missions evolve quickly. Then the defence systems might be in service for ten to thirty years.

How can armed services clearly draw the need of a precise output in such a context and accept not to change its specifications, even incrementally? PPPs are expected to overcome two of the worst biases in public-managed projects: time and costs overruns. PFI projects also face time and cost overruns, even though to a more limited extend that classical procurement programmes. If part of such drifts results from organisational inefficiencies, shared by most public administrations, it appears that specific difficulties in providing defence also explains why arms programs are uneasy to manage. The problem is intrinsically linked to the issue of defining what it is requested, so that PFI cannot appear as a *silver bullet* to overcome inherent biases.

As Allen (2001, p. 27) explains it, 'the PFI should reduce the burden on the public sector from the risks of cost overruns, provided the original contract stipulates that this risk should be transferred to the private sector contractor, thereby introducing incentives to avoid such problems'. However the targeted risk relies on financial or organisational risks. Technological or geopolitical risks are far more difficult to manage and firms do not seem to be more qualified than public services on those issues. This leads to understand how and if risks can be transferred to private contractors.

According to a 1996 report from the Adam Smith Institute,[12] the total costs of tendering for a PFI project to all potential contractors is just under 3% of expected total costs while for a traditional procurement the total costs accounts for about 1%. 'One reason for the higher cost of tendering for a PFI project', notes Allen (2001, p. 34), 'is that the time taken between offering public services projects to the private sector and the final signing of the deal can be protracted, especially for particularly intricate and technical projects'. Defence projects show high transaction costs and very long negotiations, since systems are truly specific. Defence appears as an extreme instance of the inherent costs in PFI contracting, with three consequences:

1. Specific constraints in defence requirements and uncertainty induce a high level in the total costs of tendering, exceeding those of most civilian services.
2. Such costs are bound to reduce expected savings.
3. They might become a barrier to entry, diminishing de facto competition, since incumbent firms are going to better manage contracting processes.

Moreover defence systems are characterised by strong asset specificity. These systems have very few alternative uses, and most of the time there is no second-hand market. Asset specificity constitutes a prime condition for 'hold-up' in contracting, especially in defence since most of time 'there is no alternative party with whom to contract without net costs that equal or exceed those resulting from acceding to the opportunistic contractor's demands' (Parker & Hartley, 2001, p. 8).

Actually the British MoD has signed many PFI contracts for very long periods (40 years on aircrew training simulators, 30 years for housing, 20 years for water and sewerage ...), corresponding to the life cycle of many defence systems. The problem is that long-term contracts induce three biases:

1. Suppliers tend to undertake costly investment schemes, and projects tend to have a bias towards high quality and advanced technology even though it is not the most suitable solution.
2. They create a long-term monopoly, which cannot be suitable with market mechanisms and does not provide incentives to introduce disruptive innovations during the contract duration.[13]
3. It creates a window of opportunity for hold-up strategies.

Committing in long-term contracts implies that both principal and agents may rely crucially on partnership, trust and reputation. Reputation (based on past behaviour) and trust (based on expectations of future behaviour) are central to the contracting process. This is especially true concerning defence systems, since armed forces must be sure that their partners are able to provide the requested output and are not going to go bankrupt. Defence imperatives and characteristics imply a deep cooperation between buyers and sellers, which does not correspond with the PFI philosophy – especially true competition and risk transfer.

The notion and practice of risk transfer, which lies indeed at the heart of the PFI justification, are problematic in these cases. The concept raises issues of definition and measurement that cannot be considered as trivial, particularly when we are aware that risk transfer is likely to be evaluated ex post rather that ex ante. When defence requirements concern its essential features of public good, it appears that the public sector is unable to relinquish risk and responsibility for ultimate performance – even though contracts *formally* transfer such risk and responsibility.[14]

The defence industrial base restructuring gave birth in the 1990s to few major, transnational actors.[15] One consequence of such a trend is that States face stronger and bigger private partners. It is perceptible through contract awards. 'The increasingly large scale and long-term contracts being negotiated in the name of "partnerships", remarks Shaoul (2002, p. 3), mean that public agencies can only work with a relatively small number of large corporations capable of managing and operating such contracts'. The Centre for Facilities Management[16] reported the results of a survey of contracts awarded in 1996–1998, indicating a continuing decline in the number of single service contracts awarded and a big rise in total facilities management contracting.

This transformation turns upside down the provider-client relationship, especially in defence since there are fewer firms than in many commercial activities. More and more States have to negotiate with a monopolist and competition is tough to maintain without compromising the national industrial and technological base. Even worse, States may become under influence of their private partners in their procurement choices. The PFI strategy can work when there are many potential contractors in a specific market. When the State faces an oligopoly or even a monopoly it can no longer secure a competitive bidding. Such a result is accentuated when there is a learning effect.

When conditions favour the creation of a sheltered cartel, firms have a strong temptation to collude to control the defence market, secure a high rate of profit and keep innovation under control. A relative small number of bidders leaves contractors in a potentially powerful position – reducing competition, limiting the public sector's ability to obtain value for money and affecting the nature of service provision. Moreover transferring activities to private partners might diminish armed forces' ability to evaluate output or proposed services because they no longer have in-house competencies regarding assets or services. There is a risk that the private sector dictates social and public needs because of a stronger and stronger information asymmetry.

Institutions might then become 'blind and dumb' in the medium or long term. In an innovative sector like defence this creates a jeopardising situation. Armed forces need advanced systems and if they rely entirely on private firms, these latter may propose the systems they master, not the state-of-the-art technologies and alternative systems.

The quality of public services, especially when these latter are quite specialised, relies on the expertise of public sector employees, who are not only able to describe the outputs required but know how those outputs well enough to monitor their production. One issue of moving activities to the private sector is that intellectual capital is also transferred. As embedded knowledge within public institutions is vanishing, there will be no one with the required competencies to specify services as well as judge the price and quality offered for their provision (particularly on technological issues when they are truly specific to one field).

Benefiting from Private Firms Knowledge: Towards True PPPs

In fine the government is bound to bail failing activities out, for it is the last-resort guarantor of sovereign activities. As Shaoul (2002, p. 21) puts it, 'ultimately the government cannot avoid political embarrassment and public perceptions of responsibility, and may end up footing the bill when things go wrong'. Such a situation can be uneasy in civilian services. It becomes truly perilous in the case of defence when one country's international security is jeopardised because of the bankruptcy of a crucial provider.

The dogmatic application of the PFI might lead to a travesty of the notion of risk transfer. Transfer remains apparent, and incentives toward private partners are vanishing. This is the reason why armed forces need to assess the fields for which PPPs can apply and which kind of PPP is the most relevant to a given issue. Even the British Treasury recognises that 'risk should be allocated to whoever is best able to manage it',[17] not risk transfer for its own sake. Then it seems that there is a core business of armed services that must keep under the control of State. Otherwise it could contradict armed services' own and ultimate missions.

When buyers and suppliers have imperfect information, contracts cannot be optimal in a full information sense. As Parker and Hartley (2001, p. 14) underline: 'A distinctive feature of MoD's business is the requirement to provide an operational capability in peace, crisis and war and PPP projects have to meet this criterion. This increases the difficulty of writing a long-term contract dealing with all contingencies'.

It seems indeed difficult to write contracts with complete contingent claims. Particularly when these contracts cover a lengthy period of time, technologies and costs are inherently uncertain. Such 'true uncertainty' (Davidson, 1991) leads private contractors to secure the contract by adopting a conservative perspective. Thus Kaldor (1981) showed that incumbent firms fight against disruptive technologies since these latter devaluate their assets and reduce barriers to entry. Then PPPs contracts may not adequately correspond to a provision of defence in a long-term perspective.

Furthermore, when there is a true risk transfer, one may wonder whether private contractor will have real incentives to provide the kind of innovations on which strategic superiority is based. Hart, Shleifer, and Vishny (1997, p. 1159) suggest that State ownership is superior to private ownership only when 'quality innovations are unimportant'. However this idea is relevant only under a specific definition of innovations.

While PFI helps the state benefit from financial or managerial innovations, entrepreneurship is weak or even non-existent in disruptive technologies because of the specific features of defence markets (Bellais, 2000). Firms try to manage risks by reducing true uncertainty, and 'it leads to less risky designs which are most likely to work being undertaken' (Pollitt,

2000, p. 16). Then it is very likely that private contractors do not provide the crucial, technological innovations needed by armed forces.

A restricted approach of PPPs, through PFI, can suit some basic requirements. Indeed PFI appears well adapted to services, which are not very different from civilian ones even though 'customised'. Examples listed by Parker and Hartley (2001) show such a trend: Joint Service Command and Staff College, Armed Forces Personnel Administration Agency, Defence Fixed Telecommunications Service, RAF Lyneham Sewerage, RAF Married Quarters, Aircrew training facilities (simulators).

Indeed there are very few contracts concerning defence-specific needs. The only exception is the upgrade radio communications between Command HQ and front-line submarines. One may wonder if it is worth jeopardising such sensitive fields just to know if PPPs contracts are suitable there ... PPPs must help armed services concentrate on their core task of providing operational capability and obtain better value for money from existing budgets.

The aim is to create 'true' PPPs. The PFI has been used to overcome budgetary difficulties and implement a rampant privatisation. Transferring non-core activities to private sector is relevant but there is more room for long-term cooperations between the State and private contractors. As Parker and Hartley (2001, p. 13) remark, PPPs might create 'opportunities to use spare capacity, to share overheads and to undertake long-term investment against the security of income from a long-term contract'. Such market opportunities provide support for the preservation of defence industrial and technological bases (Chang et al., 1999).

PPPs might not be limited to a public-private substitution in infrastructure investment. So-called 'innovation' mentioned in PFI literature refers most exclusively to managerial and/or financial innovations. These kinds of innovation are indeed crucial and help public administrations to provide better and cheaper services; but armed forces cannot limit their partnerships with private firms to this kind of contracting. Focusing on short-term savings may reduce the window of opportunity for technological breakthroughs and disruptive innovations, leading to a technological stalemate.

PPPs entail transferring risks associated with public service projects to the private sector. Where the private sector is deemed less able to manage the project's risks, then at least some of the responsibility must remain within the public sector. This does not mean that no partnership can be imagined in such situations. Indeed true partnerships rely on long-term grounds where each partner contributes a common or, at least, share project.

The specific features of defence needs require going beyond 'classical' PFI approaches, since the private-public relationship must be deeper than in most of civilian sectors. In such a process reputation leads to trust and reinforces the partnership. PPPs must then be assessed in a long-term perspective, as agents gain experience and learn, reducing the risk of opportunism.

CONCLUSION

While provision and production of defence have been considered as inseparable for ages, it appears that this is no longer the case on both theoretical and empirical perspectives. After two decades of pro-market policies, the place of the state evolved and much of the activities previously realised in-house have been delegated to private partners. Nevertheless outsourcing and co-development cannot indefinitely enlarge their sphere of implementation. Even though the role of public and private actors in the realm of defence has been changing since the 1990s, there are some specific activities that cannot be delegated because they correspond to the core activities of the state.

As budgets tend to decrease (in absolute or, at least, relative value), ministries of defence are looking for alternative means to keep investing. PPPs have become a preferred approach in the United Kingdom and this experiment has induced other countries to try to follow the same path. Indeed, PPPs represent a strategy to overcome limited budgets and to increase armed forces' capabilities.

Nevertheless the implementation of PPP is strongly criticised. There are several examples of mismanagement and inappropriate use of PPP, which favours an opposition to such investment approach. There is no doubt that in many cases, this alternative approach did not deliver benefits that were expected. Nevertheless this does not mean that PPPs cannot work or that they are not suitable for defence investment (Bellais & Oudot, 2009).

It appears that this tool has been misused in terms of target and implementation. This is especially true concerning defence since armed forces have in charge some essential missions related to sovereignty. Armed forces must focus on the activities for which they create the best social value and rely on partnerships when it is possible for secondary activities. The true question is to determine where to draw the frontier between defence-specific activities and quasi-generic ones.

It also appears that there is a learning curve in using effectively this kind of contracts (HM Treasury, 2011, 2012a, 2012b). Indeed, many difficulties encountered by the British MoD result from a misunderstanding of how to select a project manageable through a PPP contract and of what the benefits and limits of such partnership are – as for any kind of contract. This is the reason why a state should implement PPPs from the most simple and obvious topics for such contracts and not jump directly to complex and expensive projects, as the British MoD did with the FSTA programme.

Moreover the way PFI is implemented demonstrates that partnership might go beyond a pure market relationship. PPPs represent a dramatic opportunity to combine commercial know-how and defence assets in a 'win-win strategy'. They cannot be only considered as a means to overcome budgetary shortfalls.

Such an analysis leads to have a careful use of PPP, rather than exclude this tool from the portfolio of contractual tools that ministries of defence can use in the field of investment. PPPs provide benefits when this kind of contract is appropriately chosen and implemented. There is a question of maturity about designing and implementing such a sophisticate tool: ministries of defence need to manage the appropriate legal and technical know-how.

Beyond PPPs, such contractual innovation initiated a revolution in defence investment affairs. It helps go beyond the classical perception regarding the respective place of armed forces and private sector in defence investment and related services. The perspective of the realm of defence opens new opportunities to conceive original partnerships between armed forces and their suppliers. In fact, as the American experience shows, there are several forms of possible PPPs in defence and one should not limit the assessment of investment ways to choose between patrimonial acquisition and PPPs. There exists a wide range of partnerships, sometimes already validated in commercial or civilian activities, that armed forces should explore to improve the use of defence budgets and to reposition their resources on their core missions.

NOTES

1. An economically efficient solution is expected to reduce the well-being distortion resulting from public expenditures.

2. 'In securing new financing for public investment the PFI appears to allow government to reconcile the desire for higher capital spending with the commitment to maintaining a tight fiscal stance' (Robinson et al., 2000, p. 5).

3. PPPs form long-term contracts based on service availability, encouraging firms to minimise lifetime costs in balancing capital costs with maintenance costs. Introducing a market-oriented approach can help public sector improve its current practices, since life-cycle management is hardly known within public institutions.

4. Under the European Monetary Union, five criteria must be respected by member States: under-control inflation, low interest rate, budgetary deficit below 3% of GDP, public debt below 60% of GDP, stability of exchange rates before the creation of the euro. Directly or indirectly linked to public expenditures these criteria limit the ability of States to engage in large investment projects.

5. The 2001 FSBR highlights two fiscal rules, against which the performance of fiscal policy must be judged. Following *the golden rule*, the government will borrow only to invest and not to fund current spending. Under *the sustainable investment rule*, public sector net debt as a proportion of GDP will be held over the economic cycle at a stable and prudent level, maintaining it below 40% of GDP over the economic cycle (Allen, 2001, p. 23).

6. The Economist (2002).

7. The Economist (1995).

8. See R. Barro (1974).

9. Shaoul (2002) studies several examples of PFI contracts and shows that projects are poor value for money, very expensive, poorly appraised, and would result in service reduction without further government investment. This point of view is shared by many observers of PFI contracting in health or education services.

10. Cf. Bellais and Le Blanc (2002).

11. The Audit Commission published a 2001 report, *Building for the Future: The Management Procurement under the Private Finance Initiative*. This report emphasises the importance of specifying long-term service requirement, although acknowledged that this is often difficult. 'However, contracts should also be flexible enough to build in allowances for operating flexibility but the contracts should be clear about how changes in service initiated by the contractor will be dealt with' (Scottish Parliament Finance Committee, 2001, p. 7).

12. Butler and Stewart (1996).

13. 'The more specialised those assets, the large will be the quasi-rents at stake over that period and hence the greater the incentive for agents to attempt to influence the terms of trade through bargaining or other rent-seeking activities once the investments are in place' (Masten, 1984, p. 405).

14. Specifying that risk has been transferred to contractors is one condition to put off-balance sheet the contracts according to the British Accounting Standard Board and eventually avoid increasing public borrowing statistics.

15. The United States represents an exception, since its large domestic market helped the US preserving a truly national defence industrial and technological base.

16. Centre for Facilities Management (2000).

17. HM Treasury (1995).

REFERENCES

Allen, G. (2001, 18 December). *The private finance initiative*. London: Economic Policy and Statistics Section, House of Commons Library.

Barro, R. (1974). Are government bonds net wealth? *Journal of Political Economy*, *82*, 1095–1175.

Bellais, R. (2000). *Production d'armes et puissance des nations*. Paris: L'Harmattan.

Bellais, R., & Le Blanc, G. (2002, September). *Beyond Terrorism: Asymmetrical Threats and Defence Procurement: 'Notes pour la Réflexion Stratégique' series*. Paris: CHEAr.

Bellais, R., & Oudot, J. M. (2009). La relance des contrats de partenariat dans le domaine de la Défense en France. *Revue Française d'Administration Publique*, *130*(July), 263–274.

Butler, E., & Stewart, A. (1996). *Seize the initiative*. London: Adam Smith Institute.

Centre for Facilities Management. (2000). *UK Facilities Management Market 1999*. Strathclyde Graduate Business School, Glasgow.

Chang, I., Galing, S., Wong, C., Yee, H., Axelband, E., Onesi, M., & Horn K. (1999). *Use of public-private partnerships to meet future army needs*. Santa Monica, CA: RAND Corporation.

Cornes, R., & Sandler, T. (1996). *Theory of externalities, public goods and club goods* (2nd ed.). Cambridge: Cambridge University Press.

Davidson, P. (1991). Is probability theory relevant for uncertainty? A post Keynesian perspective. *Journal of Economic Perspectives*, *5*(1), 129–143.

Domberger, S., & Jensen, P. (1997). Contracting out by the public sector: Theory, evidence, prospects. *Oxford Review of Economic Policy*, *13*(4), 67–78.

François, P. (2000). Public services motivations as an argument for government provision. *Journal of Public Economics*, *78*(3), 275–299.

Hart, O., Shleifer, A., & Vishny, R. (1997). The proper scope of government: Theory and an application to prisons. *Quarterly Journal of Economics*, *112*(4), 1127–1161.

HM Treasury. (1995, November). *Private opportunity, public benefit: progressing the private finance initiative*. London.

HM Treasury. (2011, July). *Making savings in operational PFI contracts*. London.

HM Treasury. (2012a, December). *A new approach to public private partnerships*. London.

HM Treasury. (2012b, March). *PFI signed project list*. London.

Kaldor, M. (1981). *The Baroque Arsenal*. New York, NY: Hill and Wang.

Masten, S. (1984). The organisation of production: Evidence from the aerospace industry. *Journal of Law and Economics*, *27*, 403–417.

Middleton, N. (1998). Consistently inconsistent on private finance. *Accountancy*, (April), 89.

Parker, D., & Hartley, K. (2001). *Transaction costs, relational contracting and public-private partnerships: A case study of UK defence*. Fifth International Conference on Technology Policy and Innovation, TU Delft, Delft, 26–29 June.

Pint, E., Bondanella, J., Cave, J., Hart, R., & Keyser, D. (2001). *Public-private partnerships: Background papers for the U.S.–U.K. conference on military installations assets*. Santa Monica, CA: RAND.

Pollitt, M. (February, 2000). *The declining role of the state in infrastructure investments in the UK*. University of Cambridge: Judge Institute of Management Studies.

Robinson, P., Hawksworth, J., Broadbent, J., Laughlin, R., & Haslam, C. (2000). *The private finance initiative: Saviour, villain or irrelevance?* London: Institute for Public Policy Research.

Samuelson, A. (1954). The pure theory of public expenditures. *Review of Economics and Statistics*, *36*(4), 387–389.

Scottish Parliament Finance Committee. (2001, August). Public private partnerships and the private finance initiative: A review of recent literature. Edinburgh.

Shaoul, J. (2002, January). *The regulation of and accountability for new forms of public procurement.* Working Paper No. 17. CRC-IDPM, University of Manchester.
The Economist. (1995), Cooking the books. *The Economist*, 28 October.
The Economist. (2002), PFInancial services, *The Economist*, 14 September, p. 40.
Williamson, O. (1985). *The economics institutions of capitalism.* New York, NY: Free Press.

THE ECONOMIC AND SPATIAL SIDES OF DEFENCE SUPPORT

Josselin Droff

ABSTRACT

This chapter discusses the ongoing transformations of the French defence support. Considering the importance of economic activities related to defence support, this contribution aims at discussing the evolution of defence support and its costs for the State. The literature in defence economics presents very little analysis of defence support in its different forms. Neither space nor base locations have been deeply analysed in such a literature. We try to bridge this gap in an original research framework. We focus on the Maintenance, Repair and Overhaul (MRO) of defence platforms. More particularly, we focus on the array of measures that surround French defence support in MRO since the end of the 1990s. Considering both economic and spatial leverages, how can the organisation of MRO be optimised? With concepts from spatial economics, we propose a new look on the defence support system. We examine new economic interconnections (e.g. Public Private Partnerships and outsourcing) between military and civilian activities. More broadly speaking, this path of research could help us to better understand the new type of economic interrelations between defence organisation and 'territory' as a social fabric.

The Evolving Boundaries of Defence: An Assessment of Recent Shifts in Defence Activities
Contributions to Conflict Management, Peace Economics and Development, Volume 23, 51–73

ISSN: 1572-8323/doi:10.1108/S1572-832320140000023006

Industrialised countries typically have to cope with the challenge of public deficits. This situation theoretically involves the search for an optimal allocation of public resources. Such trend has been described as an *efficiency strategy* in public management (Metcalfe, 1984). Indeed, given that there is a constant demand for public goods, governing bodies need to create ex nihilo, reform and close some facilities, open others and generally move them from location to location in order to make them as efficient as possible to satisfy the users. In a word, the search for new forms of efficiency in public goods management seems to be unavoidable, including for military forces. Indeed, due to its high specificity (e.g. on the demand side one consumer exigent in the technical nature and reliability of the systems on the supply side many monopolistic situations due to the technical specificity of the military products and important part of the *in-house* production for strategic reasons), the Army is likely one of the latest public institutions experiencing such new public management strategies in its organisation. Furthermore, we suggest this huge process has evolved at an increasingly rapid rate since the beginning of the 1990s. Our assessment calls for new questions about how to set up the production of the public service of defence and how to define a new research agenda in defence economics. The focus of research we have chosen is the topic related to the economic organisation of the Army, with a focus on the spatial dimensions and the costs associated with them. More particularly we study defence support side of the Army with example taken from the Maintenance, Repair and Overhaul (MRO).

Military functions are divided into *operational functions* or *support functions*. Actually, considering such a dichotomy, defence support for military systems can be considered as a complex service delivered to military forces. We define defence support as all the technical, material, financial and human means contributing to the readiness of military forces. A recent conception of defence production consists in sending military forces on missions and preserving their ability to project. Considering this definition and for a given endowment in military platforms and soldiers, we call defence support supply chain all the activities upstream of defence production. Then, we call dynamics of efficiency in defence support the trend in the minimisation of cost in the whole defence support under given constraints. These constraints include economic constraints but also non-economic like the readiness of armed forces, national security or quality and users satisfaction with the delivered service. Indeed, for given military forces with their equipments in a given area, optimality is theoretically the economic and spatial structure minimising the cost of defence support. The main objective for the public planner becomes to match up defence support supply to the military forces demand in the area. The chapter deals more specifically with the maintenance of military systems on the national territory, but we have to keep in mind that the MRO processes is part of the whole defence support, that is defence support supply chain. Such a complete chain includes other defence support functions that we do not treat here, for example general defence support (GDS) (e.g. food delivery) or some kinds of specialised defence support (SDS) (e.g. maintenance of infrastructures).

As defence establishments are often an important part of a local economy of a region, the literature in defence economics has mainly focused on assessing the economic impact of military bases through direct, indirect and induced impacts on employment and income (see e.g. of static empirical models, Asteris, Grainger, Clark, & Jaffry, 2007; Leontief, Alison, Karen, David, & Edward, 1965; Rioux & Schofield, 1990). Moreover, if we examine the literature about defence in regional economics (Breheny, 1988) or one of the most famous references in defence economics (Hartley & Sandler, 1995, 2007), very little is said about defence support facilities in their various forms. Although several econometric methods *spatialise* defence activities (Andersson, Lundberg, & Sjostrom, 2007; Erickson, 1977; Paloyo, Vance, & Vorell, 2010), space and base concepts have not been deeply analysed in this kind of literature. We try to bridge this gap by defining and discussing an original research framework. Several factors are involved in the organisation of defence support and its cost for a country. If some of them are related to economic activities in general (e.g. budget constraint, cost of inputs or the nature of the technology), others represent more specific constraints due to the military nature of defence support activities (e.g. strategic choice in location, strong importance of the technology-driven support, ageing of military platforms, increased involvement in conflict and military operations and its consequences on the use of systems). Considering both economic and spatial leverages, how can the cost of attaining defence capabilities be minimised under the previous constraints? Keeping in mind the cost of defence support and specific constraints associated with it, how should we organise the *spatialisation* of defence support activities? This chapter aims at giving some first elements of answers to these questions.

The chapter reviews in two sections the broad changes at work in the support of French military equipment. From the technical readiness of the main military systems and cost trends in defence support since the end of the 1990s, it presents a review of factors that reshape the needs of defence support and do influence its cost (section 'Readiness of Armed Forces and

Trends in the Cost of Defence Support'). It continues with an overview of the both economic and spatial leverages that the public planner sets up to raise readiness and reduce the cost of defence support (section 'Transforming Defence Support: Reducing the Cost and Raising the Readiness'). In the conclusion, it makes preliminary proposals for further research towards an optimal organisation of defence support related to *spatial economics.*

READINESS OF ARMED FORCES AND TRENDS IN THE COST OF DEFENCE SUPPORT

We first comment on the evolution of cost in the specific defence support (MRO) and the trends in readiness of French military systems since the end of the 1990s (section 'Readiness of Armed Forces and Evolutions in the Cost of Support'). More precisely, considering the readiness we try to draw a synthetic picture of the French military equipment readiness connected to the MRO process and its evolution. Then we explain these evolutions with both cyclical and structural factors (section 'Assessing the French Situation: The Causes').

Readiness of Armed Forces and Evolutions in the Cost of Support

The MRO Process

MRO is a very specific type of defence support. It aims at the readiness of military systems, including weapon systems with the best cost-effectiveness relation. It deals with the obsolescence of systems, including their retrofit and specific support for ageing platforms. The MRO process is both an economics and logistics problem par excellence, that is the cost minimisation of attaining stipulated capabilities (Menderhausen, 1958). In 2004, 13% of the employment of the French Ministry of Defence was allocated to such a defence support (Cour des Comptes, 2004). In 2009, the whole MRO cost was estimated to be about 15% of the French Defence budget (without retiree pensions) and about 27% of the investment expenditure. Moreover, the MRO costs of military systems tend to increase in the long run for several reasons, which we detail further in the chapter. As we see in Fig. 1 with the example of the French Air force fleet (fighter), MRO costs

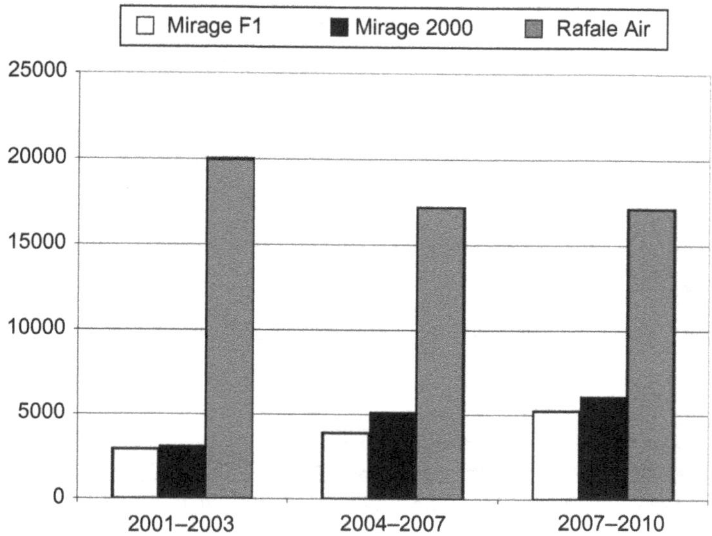

Fig. 1. Planned Maintenance of Equipment (PME) Credits *per* Hour of Flight in Euros, for Three Platforms in the French Air Force. *Source*: Author, data from French Ministry of Defence.

of modern weapon systems tend to increase in time for ageing platforms and from generation to the next.

Moreover, data from the *Congressional Budget Office* (CBO) (2011) give us an overview of the escalating maintenance and support costs in the case of US Army (see Fig. 2).

Assessing the French Situation: The Trends in the Readiness

We first assume that readiness rates are basic and commonly indicators for measuring the effectiveness of MRO. Using public reports, we have rebuilt average technical *domestic readiness trends* for each main military fleet. Domestic readiness is different from *in-theatre readiness*, which is generally between 90% and 95%. However, assessing the domestic readiness rate is important since it is under condition of the *in-theatre readiness* and, for a given budget, does influence the training level of troops and the future potential of the Army. At the end of the 1990s an important breakdown affected the domestic readiness rates of the French military systems. The three armed forces were affected, but in different proportions. We now present a service-by-service assessment of the readiness trends.

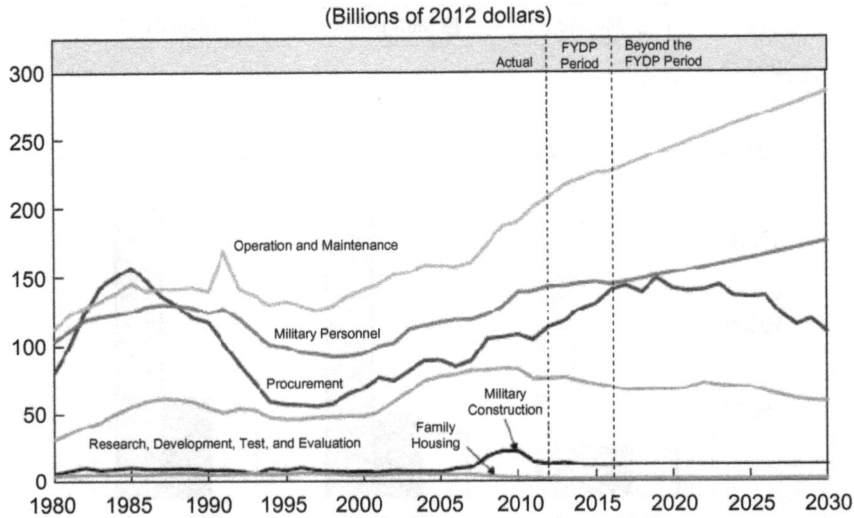

Fig. 2. CBO Projection of Base-Budget Costs of DoD's Plan by Type of Spending. *Source*: CBO (2011, p. 7). *Notes*: Base-budget data include supplemental and emergency funding before 2002. FYDP = Future Years Defense Program; FYDP period = 2012 to 2016, the years for which the Department of Defense's (DoD's) plans are fully specified.

The French Navy. The average technical readiness rate of the French Navy fleet lost more than 30 points of readiness during the 1997–2000 period while the fleet was smaller and smaller (110 ships in 1990, 82 in 2002 and 70 in 2010, probably 60 in 2020) (Meyer, 2002). Moreover, logistic functions and infrastructures of the Navy deteriorated during this period (Freville, 2005). Because of its specificity the trend does not affect the nuclear submarine fleet, which is dedicated to deterrence policy and whose readiness technically equals 100% (by definition of the deterrence policy).

The French Air Force. A long decreasing trend in the readiness of French military aircraft is noticed from 1996 until today (Freville, 2008). The decrease is especially pronounced from 1996 to 2001. Then the readiness increased during the first half of the decade from 50% of technical readiness in 2000 to 60% in 2004. The deterioration in the readiness repeated again and the steady-state readiness rate of about 60% of the 2003–2006 period started to decrease from 2006 to 2008. In 2011, Air Force readiness was below 60%. But, as the Air Force has a heterogeneous fleet, the overall trend of readiness should be combined with a full picture of the readiness for each type of platform. Indeed, to be fully studied, the readiness needs a case-by-case study. The diversity (in age, use and technology) of fleets is particularly a relevant factor to analyse the MRO processes of the Air Force, the technical readiness and the time of training associated with for pilots. For example, the Air Force maintains very old tanker aircraft *KC-135* (47 years), old transport aircraft like *C-130* (30 years old) or *C-160* (40 years old) along with the much newer *Rafale* aircraft (6 years old). As a result a global trend of the whole aircraft fleet should be combined with a full picture of the readiness of each fleet (or better each kind of aircraft) during the analysed period. Finally, having in mind the previous observations, combat aircraft and helicopters are able to reach the targeted time of flight required by Air Force's standards in the 2000s, which is not the case for transport aircraft (Dulait & Carrère, 2008).

The French Army. In the 2000s, the French Army experienced a substantial drop in the readiness of its main systems. For instance, light and heavy armoured tanks experienced an important decrease in readiness in the decade. The ratio is particularly low for the heavy *Leclerc* tank whose readiness was under 40% in 2008. Another example is that of Army helicopters whose readiness shortfalls were particularly extreme at the beginning of the decade 2000 (the army operating about 70% of the French helicopters). Moreover this downward trend coincides with a lack of practice for pilots. The Senate noted that French Army helicopter pilots realised 250 hours of flight per pilot in 1992 and 140 hours of flight in 2002 (with a North Atlantic Treaty Organization (NATO) standard of 180 hours of flight per pilot) (Vincon, 2002). However, this trend reversed at the end of the 2000s when pilots were able to achieve both NATO and French standards in 2009 (Dulait & Carrère, 2008).

Assessing the French Situation: The Causes

The causes in the decreasing readiness combined with an increasing cost of the MRO are multiple. We have to keep in mind that readiness is economically related to MRO cost, and a financial trade-off exists. That trade-off depends on both budget and MRO cost. Moreover, assessing the complete MRO cost is a difficult exercise because of both scarcity of data and cross-domain characteristic of the function (MRO cost involves manpower cost, weapon maintenance cost, engine cost, spare parts, outsourced expenditures for maintenance, infrastructure costs). However, drawing a typology of causes in such a nexus between cost and readiness is possible. On the one hand, cyclical factors play a role, for example a decrease in financial resources or the increasing cost of raw materials. On the other hand, the cost of defence support is function of structural factors, such as the increasing share of technology intensive systems or the organisational models of production. More precisely, for the three branches of the armed forces, we identify six distinct different factors that do influence the trends in readiness, trends in cost or the relation between readiness and cost.

Cyclical Factors Involved in the Decrease of Readiness and Increasing Cost of MRO

Budgetary Resources. The level of the PME credits gives a first look on the trend of budgetary resources allocated to the MRO process. After an average increase of about 2% between 1977 and 1993, a strong average decrease of almost 20% is noted for the 1993–2002 period. Indeed, considering the trend at the end of the 1990s, the main cause of the decrease in the readiness of military platforms was first analysed as a consequence of a decreasing trend in maintenance credits (PME credits) (see Fig. 3). During

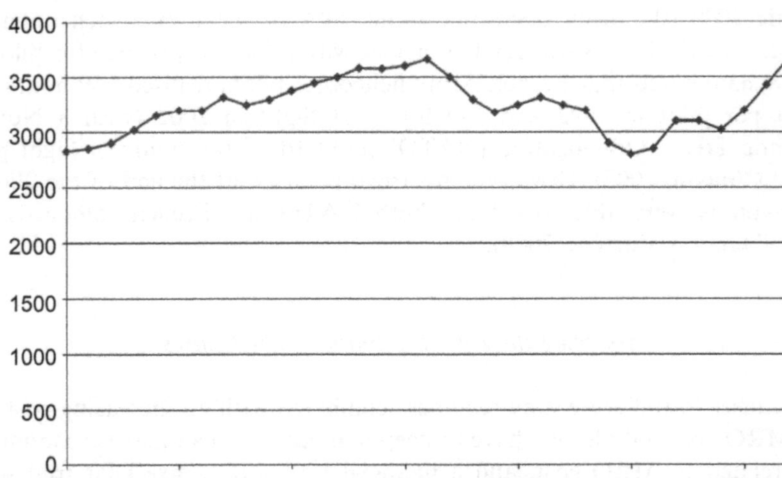

Fig. 3. PME Credits Trend 1977–2009, in Millions of Euros. *Source*: Dulait and Carrère (2008), p. 42.

this period, PME credits were considered as an adjustment part of defence expenditures with a defence budget under constraint and a constant increasing part of operating expenditures (later due to the importance of workforce expenditures in the transition towards an all-volunteers force). The consequences of such a trade-off between PME credits and operating expenditure credits were called the *budget notch* in PME credits.

Increasing Complexity in MRO Process and Its Consequences on the Cost of Inputs. The cost of inputs influences the total cost of MRO (e.g. raw materials, workforce, spare parts).

First, in the case of modern and highly technological equipment, the price of inputs like composite elements in modern weapon systems or on-board electronic systems (Hebert, 1995) can be an important driver of total MRO cost. For instance, the share of parts that are made of composite elements has risen from 7% for a *Mirage 2000* to 26% for the *Rafale* (Deo, Starnes, & Holzwarth, 2001). Such composite elements enhance performance of aircraft and theoretically reduce the number of MRO operation. However, they also involve very specific and costly assets and might suffer from bottleneck in supply considering the increasing demand for such elements in various manufacturing sectors. Second, workforce is an important element of the increasing cost of MRO. This part of the cost attributed to the *workforce element* has two sides. The first is a qualitative one and has to be found in the increasing complexity in MRO tasks and the skills necessary to deal with more and more complex military systems. Such more and more complex military systems need highly MRO skills that could hardly be found in the draftees or inside the Army. The second is a quantitative one. With the professionalisation of the Army, some basic but numerous tasks, formerly executed by the draftees (costless workforce) are now conducted by professionals (military or civilian costly workforce). Finally, the increasing cost of spare parts and their management incites the Army to use its stocks and to develop a kind of *aircraft cannibalism*.

Reforms, Institutional Changes and Historical Events. We argue that reforms, institutional changes and historical events do reveal and possibly impact the cost structure of MRO processes with *hysteresis* effects. First as an inheritance of an Army organisation's model made for the Cold War, the location of military bases (mostly with their own MRO infrastructures) and units settlements (especially for the Army) were located in the North-East of France. Moreover, the stocks of spare parts were quite over-dimensioned and often located just near from weapon systems. For

20 years, both the decrease in the size of the Army and the organisation oriented towards a *force projection* model have involved a change in the management of the MRO process especially in its spatial side (e.g. warehouse location, Army fleet's settlements in order to be shipped). As a result, some *military bases* have become more important than in the past for the French military system. To give a concrete example, the ports of Toulon and Marseille in the South of France have become the first force projection port for the French Army, combining the Navy projection fleet with the proximity of the big Canjuers Army base. Second, institutional processes, resulting from the setting up of regular Army in 1997 or the transition from a public industry (i.e. arsenal organisation) to a private one impact boundaries and do have mid-term consequences on the cost structure. They do reveal the real cost of MRO and possibly increase the cost of MRO processes with *transition costs*. For instance, the French Army has experienced a new look on its MRO cost due to the transition of *Groupement Industriel des Armements Terrestres (GIAT)* Public industry to the *Nexter Group* (assessment of the spare parts, reorganisation of the supply chain on the national territory and social measures due to a decrease in the workforce dedicated to military production). Then, the French Navy experienced a similar situation with the transition from *Direction des Constructions Navales (DCN)* to *DCNS group*. Such big changes call for a consideration of the *path dependency* in the management process of MRO (in logistics, skills, communication at work, spatial organisation and history of facilities settlement, etc.).

The Structural Factors Impacting on the Readiness and MRO Cost of Military Systems

Ageing Systems and Obsolescence. The Army copes with budget constraints and tends to extend the lifecycle of military systems in place of replace them. As a result, nowadays, most of military systems are engineered to have a very long life. Such an increase in the length of their lifecycle can be an important cause of the increasing cost of MRO. As a result, the trade-off between increasing the lifecycle of a system or replacing it does impact the cost as ageing of platforms is correlated to an augmentation of breakdowns. Fig. 4 is taken from Pyles (1999) and shows the heavy-maintenance workload ratio for several transport aircraft of the United States Air Force (USAF). More precisely, the figure depicts the growth rate for a range of average fleet ages. The rate is expressed as a ratio of heavy-maintenance workload over time to the workload at the first heavy-maintenance inspection. Moreover, the weight of the system cost (MRO, retrofit) has risen

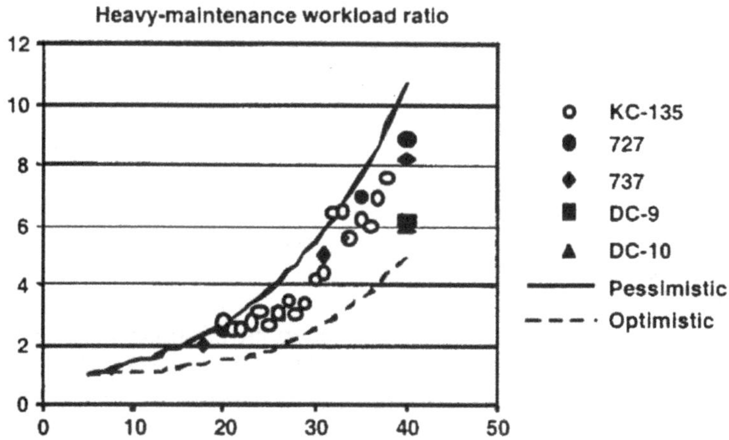

Fig. 4. Heavy-Maintenance Workload for Large Aircraft over Time. *Source:* Pyles, RAND Corporation (1999, p. 4).

during the lifecycle of the platforms. In the French case, some military transport fleets are getting older. For instance, in the French Air Force, some tanker aircraft are more than 45 years old, transport helicopters are about 30 years old and transport aircraft and *Super Etendard* fighter aircraft about 30 years. In the Army light armored vehicles Véhicule de l'Avant Blindé (VAB) are more than 25 years average old with an increasing deployment and light gunship helicopters are about 28 years old. The French Navy maintains antisubmarine frigates that are about 30 years old and operates an ageing logistic ships fleet (e.g. oil tanker, repair ship). Ageing fleets impact significantly the cost of MRO. Considering such a trend, the maintenance is more and more planned in the long run with a search for optimisation during the lifecycle of systems.

More and More Technology Intensive Systems. The modern defence systems separate the platform and a complex combination of electronic systems typically pull together, including defence, attack and security systems. These systems are often called *systems of systems* (Gates, 2004). Both procurement and maintenance costs of such systems are higher than the cost of simpler weapons. Procurement costs have been studied and increase with an annual average rate of 7% (Hartley & Sandler, 2003). Maintenance costs have also an important part in the overall lifecycle cost of military systems. Complex systems of systems cannot be maintained with generic technologies and even less with generic tools. On the contrary, they need highly specific tools and spare parts (e.g. composite spare parts). They also need different skills very close from engineering skills. Their MRO time process is likely correlated with their complexity. Finally, all the conditions above cannot be gathered without specific and expensive infrastructures. As a result, we assume the increasing technology and specific assets associated with its management to be a major cause in the increasing cost of MRO. To give a concrete example, the cost per hour of flight of the *Caracal* (medium helicopter gunship) is about 10 times what it is for the *Gazelle* (light helicopter gunship). Moreover, in 2004, the yearly average MRO cost of the heavy *Leclerc* tank was about 2.7 times more expensive than the same type of cost for the *AMX 30 B2* heavy tank (Marini, 2004). Finally, for a given budget constraint, the increasing cost of technology is typically combined with a decrease in fleet size. Consequently, such a phenomenon induces less perspective of economies of scales (mainly economies on fixed cost and learning economies in MRO processes). This is especially true for countries with small military fleets.

Use and Overuse of Military Systems. Finally, the increase of missions in operation is another important reason for the deterioration of the readiness of military systems. Such operations require a more intensive use of platforms and therefore accelerate their ageing. Since the first Gulf War, in rapid succession, foreign French interventions have particularly solicited military platforms (mostly air and ground equipments). Previously a cyclical factor in the evolution of MRO cost, the use and overuse phenomenon has become a structural one. Moreover, in-theatre operations have to be financed per se and do impact on financial trade-off, and finally on the resources available to the whole defence support. We have to keep in mind that all of the factors we mentioned are constraints in the resource allocation dedicated to the armed forces, including defence support. All of them might combine to create a complex increasing MRO cost spiral with consequences on readiness. For instance, MRO process combines day-to-day maintenance and obsolescence problems with increasing demand linked to technology-driven systems. Such a problem we call the *New&Old problem* is partly responsible for the increasing cost of MRO in the long run. Furthermore, between all of these causes, there is a kind of *glue*. More precisely, the organisation of the MRO process is this *glue* and do influence its cost effectiveness. To improve this cost effectiveness, the public planner can activate economic and spatial leverages. We now present these leverages.

TRANSFORMING DEFENCE SUPPORT: REDUCING THE COST AND RAISING THE READINESS

We believe that the problem of the cost of defence support has to be conceptualised from both economic and spatial perspectives. Hence, we present economic and spatial leverages used by the public planner in the management of defence support. *Economic leverages* are all the means contributing to raise the cost effectiveness considering the internal functioning in defence support system (e.g. way of resources allocation, changes in management processes, changes in contracting processes) (section 'Economic Leverages in the Reorganisation of Defence Support'). But, if the cost of defence support has a monetary side, it also has a spatial one. Hence, we call *spatial leverages* all the means contributing to reduce the cost of defence support with leverage on the location of defence support facilities and their belonging network (section 'Spatial Leverages in the Reorganisation of Defence Support'). We particularly would like to address the importance of concepts in the core of spatial economics. These concepts are economies of scale and externalities (pecuniary and technological externalities). Our *leverage*s' interact with each others making spatial consequences of economic measures and economic consequences with spatial causes. As a result, the logistic dimension becomes really important in our framework.

Economic Leverages in the Reorganisation of Defence Support

Resource Sharing and Increase in the MRO Credits

In Fig. 5, the decreasing trend in PME credits is more pronounced when compared to defence investment credits. The ratio increased smoothly at the end of the 2000s (at a much later date after the first warming in the reports at the beginning of that decade). We argue such a lateness to reverse the trend strengthened the financial effect in the *readiness crisis*. The current trend seems to be in support of increasing budgetary resources for MRO. For instance, in 2009, MRO credits allocated to the aeronautic support increased about 10% compared to 2008 (Dulait & Carrère, 2008). Such a decision was made to cope with the increasing cost of highly technological systems as the heavy helicopter *Tigre* and its specific infrastructure in the Army for instance. However, maintenance credits are only the financial side of the economic leverages able to be used by the public planner to optimise defence support organisation.

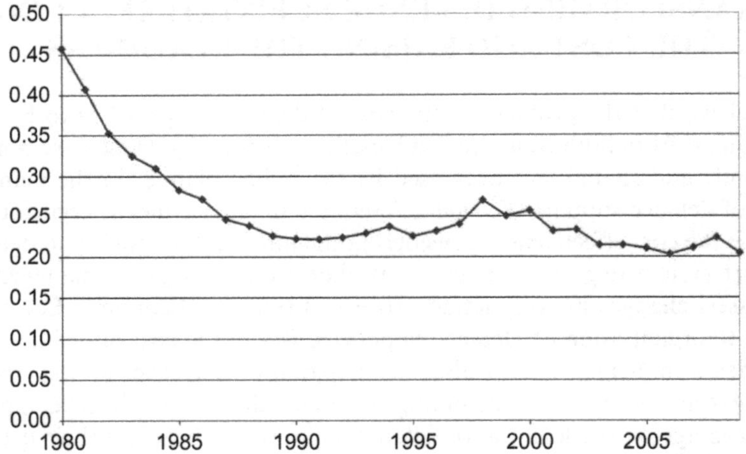

Fig. 5. The Trend in PME/Investment Expenditure Ratio 1980–2009. *Source*: French Ministry of Defence – 2011.

The Outsourcing of Defence Support Activities

To outsource means to attribute to a private partner a part or the totality of a function before provided in house. Some of defence support activities are sometimes contestable and can be separated from the whole defence supply chain in a kind of *unbundling* movement. Going further in such a movement related to labour division and technology of production, these activities could be split off into autonomous entities (public or privately owned). These entities may be subjected to competition, while leaving the natural monopoly military support chain. Theoretically, if competition exists, military outsourcing provides opportunities for savings. Moreover, in a Smithian vision of work process with a finer labour division, that is a greater specialisation in tasks, outsourcing should theoretically improve the quality of support services.

As Fig. 6 shows, French defence outsourcing expenditures have notably increased during the decade 2000. This trend is best viewed over two distinct periods. A first period from 2000 to 2005 is mostly due to an increasing trend in workforce expenditures (because of the French armed forces' professionalisation). In a second period from 2006 to 2008, outsourcing expenditures rose by 76% in value. Though difficult to isolate, an important part is attributable to defence support, especially MRO expenditures, foreign operations expenditures and to a lesser extent a rise in the GDS due

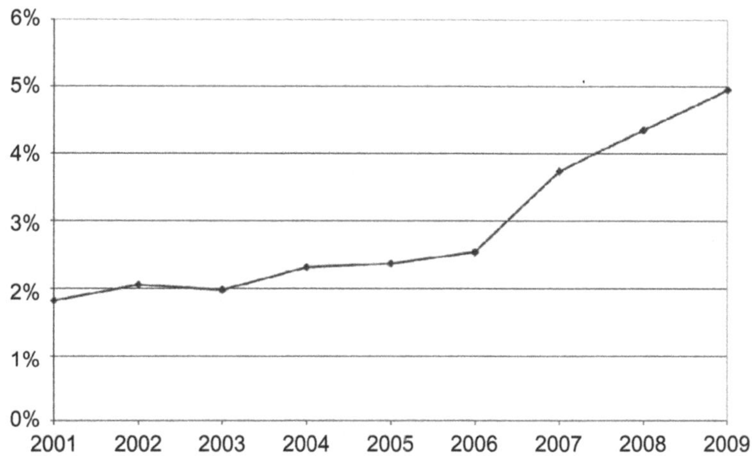

Fig. 6. Outsourcing in Defence Expenditures 2001–2009. *Source*: French Ministry of Defence – Observatoire Économique de la Défense (OED).

to the ongoing reform of the French Defence Bases. The new reform or Defence Base reform aims at rationalising GDS functions as administration, housing maintenance in a given perimeter, a joint-army perspective and for a unique administrative body. This ongoing reform experiments or reinforces outsourcing strategies in transports, reprography, garden maintenance, light vehicle repairing and maintaining, maintenance of heating systems, etc.

In the next three paragraphs, we offer three concrete examples illustrating this outsourcing strategy:

First, in the transport field, the low readiness rate of the transport aircraft combined with the age (hence the MRO cost) and the needs to transport troops, lead the Army to outsource a part of the logistic transport in foreign theatres (e.g. transport of helicopter gunships, armored vehicle, etc.). Considering such an outsourcing strategy in logistics in foreign theatres, only about 12% of the *C-160 Transall* and 14% of the *C-130 Hercule* were used in logistics mission (Marty, Sordi, & Viollet, 2008).

Second, in the MRO field, since 2007, the Air Force base N-709 of Cognac has outsourced MRO processes, pilot training and the management of the training fleet. In this case, good results have been noted with an increase in readiness (more than 90% combined with a roughly estimated decrease in cost about 35%) (Marty et al., 2008).

Third, another interesting case is given with the school-base of Dax where the French Army no longer manages the training of its helicopter pilots, just buying flight hours and benefitting from 28 modern helicopters *EC-120* available daily. This solution has increased the readiness of the operational fleets and offered to the Army an opportunity of reallocating highly specialised employment in MRO (for very specific and complex tasks in other sites or on other fleets). Moreover, in this case, a decrease about 8% in cost has been estimated (Carrez, 2010). In the outsourcing field, the search for an optimal path in the supply of defence contributes to reshape *new boundaries between state and markets* (Bellais, 2004).

New Management Models

New structures and new management models also contribute to raise readiness and to reduce the cost of defence support (Cour des Comptes, 2004). These new structures adopted a MRO based on fleets (1) and process oriented towards the natural background of systems (2). To simplify, one deals with naval systems, another with air systems and the last with land systems. As a result, the French ministry of defence created specialised joint-army structures to optimise the maintenance with *demand processes oriented.* The setting of those new administrative bodies started in 2000 with the creation of the SSF (Navy Fleet Support Structure). At the same time, the SIMMAD (Structure of Integrated Aeronautics Management) was created to manage the MRO aeronautic supply chain (for the three services). Finally, the French ministry of defence set up the SIMMT (Structure of Integrated Management of Ground Equipment) in 2010. These new administrative bodies permitted the ministry to gather all the skills and know-how under one and only one authority, with the possibility of new and possibly more efficient forms of partnerships. For instance, in an agreement between *Safran* and the French ministry of defence, a pooling of resources is intended to smooth out the peak and troughs of maintenance work on military engines for the French services and *Snecma*'s foreign clients.

Such new administrative bodies have also developed and managed the *in service support* of MRO (often outsourced) with *global contracts*. For example, the Air Force acquires a readiness time of its aircraft that is ex ante stipulated in the contract with a private partner. Another example is given with a contract signed in 2008 for the maintenance of light armoured vehicles. In this contract, *Nexter* (French firm specialised in the engineering of ground military platforms) certifies a readiness rate of 95% in all operation theatres combined with the maintenance of all the vehicles and the management of their spare parts on the French territory (Dulait & Carrère, 2008). These new ways of managing the MRO supply chain seemed to have led to a decrease in the MRO cost, combined with an increase in readiness. For example, both savings about 20% in the naval MRO and increased in the readiness of the fleet were noted during the period between 2003 and 2008 (Adam, Beaudouin, & Fromion, 2009).

Last, these new structures also adopt a lifecycle view involving all the actors of the *military supply chain* and search for reducing the cost of the different parts in the carrying cost of weapon systems (including disassembly and destruction costs) (Meyer, 2002). Such an approach leads to identify potential cost items early in the weapon system design phase (e.g. reducing obsolescence risk with standardisation methods) and therefore to develop proactive and reactive initiatives.

Spatial Leverages in the Reorganisation of Defence Support

Economies of Scale, Size of Defence Facilities and 'Market Power'

Economies of scale are considered to be internal to productive units and result from a decrease in production cost whereas production increases in volume. In fact, *there are economies of scale if a small proportional increase in the levels of all input factors can lead to more than proportional increases in the level of output produced* (Panzard & Willig, 1977). Such economies are often associated with the increasing size of productive units and with mass production (Fujita & Thisse, 1997). In the case of defence support, such *economies of scale* are related to indivisibilities and have three possible origins.

First, in the short run (when fixed costs are given) they are due to a wider or better sharing of variable inputs and a split of fixed costs (e.g. overheads costs) over a larger base. In our case, some military infrastructures (administrative entities, warehouses, depots, maintenance specific infrastructures, etc.) are indivisible, and their cost needs to be shared to lower the fixed cost per unit produced. As a result, to better share such fixed costs among armed forces, the public planner can set up joint armed forces policies in different support fields. It finally results a division of fixed costs associated with indivisible assets as much as they can among military units.

Second, economies of scale can be enforced with a specialisation of the workforce and the improvement in production processes due to the learning effect between workers. Such *learning economies* are basically the same as in the labour specialisation in the example of the *pin factory* of Adam Smith (1776).

Third, we note economies that are not merely pure *economies of scale*, but consequences of a bigger size. An increase in the size of defence support facilities possibly concentrates buyers in a given area and gives them more market power and may help in price negotiation. Such a trend in the increasing size of the market power is enhanced with *guiding contracts* defined at the central level and managed by the new structures we mentioned before.

The spatial trend in the armed forces geography could be *less military bases but bigger bases*. Such a movement is a concentration one. From 2009 to 2016, 83 French military sites have been or will be closed and 33 will be moved (on a total of 471 sites). According to us, this kind of trend is in line with Krugman's (1998) argument: *Economies of scale do enforce the geographic concentration of activities*. Under the influence of economic forces (mainly budget constraints, increasing cost in the maintenance of military systems and increasing cost in other defence supports) some military sites have increased in size, in the search for economies of scale. However, if the size depends on economies of scale, it is also limited with transport costs and coordination costs between the shared indivisibility and dispersed military units. As a result, the size of defence support facilities could be determined with a trade-off between economies of scale and transport costs (Fujita & Thisse, 2003). Moreover, as the defence organisation is not merely the result of an economic process, the equilibrium between concentration and dispersion on the national territory should be analysed with other non-economic forces (e.g. national doctrine for the coverage of the territory including operational and strategic considerations, local lobbying and *hysteresis* effects resulting from history).

Spatial Agglomeration of Activities and the Search for Agglomeration Economies

Agglomeration is the concentration of both producers and consumers in a single location. In the search for localised increasing returns, agglomeration economies are considered to be external to economic agents (Catin, 1991). The microeconomic reasons for agglomeration effects are complex but the literature sums it up in three main reasons.

First, Marshallian externalities enhance the productivity of agglomerated agents (mostly firms), which benefit from each others from the fact to their collocation. Concretely, they can exchange information, especially tacit information that can reduce their coordination costs and finally raise their individual productivity. Furthermore, the location of some defence facilities may influence the location decisions of firms that wish to locate near prime contractors, or the expansion decisions of firms currently located near contractors. Moreover agglomeration induces both a better specialisation and adjustment of the workforce on local labour markets. Such an effect would be enhanced with the increasing size of the market, that is possibly with the size of a city related to a given defence base.

Second, an increase in the size of the market (resulting both from the increasing size of a given production unit and from the agglomeration of economic agents) develops the possibility for producers to sell more intermediate or final inputs. Such localised (and non-tradable, e.g. maintenance services), highly specialised and diversified inputs can lead to an increase of the whole productivity of the agglomeration (Abdel-Rahman & Fujita, 1990). Moreover, the literature has shown that inputs as financial or maintenance services do positively influence local growth as (Hansen, 1990) shows. Latter effect incites to enlarge the concept of *spin-off* effects defined as technological transfers to the local civilian economy *via* new product or production processes. Such technology transfers reduce business costs and improve productivity with consequent effects on real personal income (Weston & Gummet, 1987).

Third, with the possibly increasing size of cities related to bigger defence bases, the opportunities to share possibly common infrastructures to reduce the fixed costs could become interesting for military and civilian. We call that the *territorial duality* between military and civil economic activities in the evolving boundaries of defence. In urban economics, for a given economic organisation, such a phenomenon related to the size of the city and to the density of agent is well known as *urbanisation economies* (Jacobs, 1969). Indeed, to reduce the cost of defence support, the used technology should be examined with a look on the possibilities of sharing dual indivisibilities in production with the civil sector.

Following the literature, we assume the emergence of agglomeration economies in increasing returns to be a consequence of indivisibilities in the place where economic agents tend to agglomerate. They mainly take action in a mechanism of *learning, sharing and matching* identified by spatial economics theorists. In our case, an increase in the size of military bases could lead to agglomeration economies with both the concentration of highly specialised maintenance activities and the multiplication of links with other activities such as civil activities for instance (including transport infrastructures). Furthermore, such clusters resulting from the evolution of the traditional boundaries between the activities (e.g. civil & military boundary or public & private boundary), in spite of the classical *spin off* effects could lead to more complex reciprocal *spill over* effects between military economies and their surrounding civil economies.

Optimisation in Logistic Flows

Finally, from the two other spatial leverages, we assume that the logistics, as the management of flows in goods, services but also information is in the core of such spatial leverages. That is nothing new in the military science, which is used to optimising flows between units for a long time. However, the importance of logistics we are talking about is more focused on the national territory in a peace context (1), with a cost-effectiveness perspective (which is different from the traditional military effectiveness perspective) (2) and possibly civil-oriented or dual-oriented (which is different from an in-home optimisation) (3).

CONCLUSION: TOWARDS A GENERAL FRAMEWORK FOR AN OPTIMAL ORGANISATION OF DEFENCE SUPPORT

In this chapter, we have taken first steps towards a theoretical framework to better understand the organisation of the specific defence support with both economic and spatial perspectives. We have examined how the public planner has contributed to raise readiness and lower defence support cost with economic and spatial leverages. In the coming years, defence support in its various forms and functions will probably be one of the best economic interconnections between military and civilian activities. Often analysed in R&D activities, such interconnections are often called *productive dualism*. For economic reasons due to the cost of defence support, this *duality* seems to become the rule and reshapes the boundaries between military and civil activities (Bellais, 2004). The analysis presented in this chapter leads to additional questions: what will be the shape and functions of the defence bases of the future? Considering the nature of defence support activities and its *unbundling* movement, where should defence facilities and tasks associated with be located?

First, the proper location of a given defence support facility depends on the nature of the support needs and more particularly on the technology used. Following (Baldwin & Venables, 2010), *we take seriously the fact that technology engineering of the product dictates the way in which different stages of production fit together*. Beyond the production process, a similar analysis can be conducted about the support after those systems have entered into service and about their possible evolutions throughout their in-service phase. Considering the importance of performance related to the technology and the associated cost, such focus on the analysis of location is particularly relevant in the economic analysis of defence support chain.

Second, it could be more relevant to focus the economic analysis of defence support on *tasks* than *activities*. Indeed, the notion of *tasks* seems to be easier to connect with the notion of *skills* in a more advanced labour division in the field of defence support field. Moreover, with this notion of *task*, the outsourced functions of defence support functions might be more interesting to examine in a globalised context.

Third, as regards as the raising weight of technology in defence systems, our research suggests that speaking about *territorial defence system* in the organisation of defence support is much more relevant than speaking about *military base*. The *territorial defence system* could hierarchically be built with a limited number of sites specialised in GDS or SDS and an even more limited number of sites highly specialised in MRO. As well as being in the right place (i.e. the optimal place as regards as the technology used in the production process for a given defence support service and a given area), all these sites should have the well-fit size to experiment increasing returns, mainly economies of scale (1), agglomeration economies between military units and the civil sector (Marshallian localisation economies and Jacob's urbanisation economies) (2). Such suggestions would lead to think about an optimal number of sites defence support system needs, which can be financially affordable for a country.

All these preliminary elements could be used as another starting point for further research. All the elements and especially the *leverages* we mentioned above should be carefully examined if the public planner sets up a strategic realignment of the military bases or MRO sites (Van Roo et al., 2011). It could lead up to go beyond the current cost-sum analysis associated with the base closures (for instance in the Base Realignment and Closure (BRAC) process in the United States). It could also leads to look for long-run public management perspectives in order to search for the reliability of an optimal number of defence installations and capabilities; especially those considered as critical interconnected infrastructures.

REFERENCES

Abdel-Rahman, H., & Fujita, M. (1990). Product variety, Marshallian externalities, and city sizes. *Journal of Regional Science*, *30*(2), 165–183.

Adam, P., Beaudouin, P., & Fromion, Y. (2009). *Rapport N°1378 par la commission nationale de la Défense et des forces armées sur l'exécution de la loi de programmation militaire pour les années 2003 à 2008*, Assemblée Nationale, 95 p.

Andersson, L., Lundberg, J., & Sjostrom, M. (2007). Regional effects of military base closures: The case of Sweden. *Defence and Peace Economics*, *18*(1), 87–97.

Asteris, M., Grainger, J., Clark, D., & Jaffry, S. (2007, February). Analysing defence dependency: The impact of the royal navy on a sub-regional economy. *Defence and Peace Economics*, *18*(1), 53–73.

Baldwin, R., & Venables, A. (2010). *Relocating the value chain: Off-shoring and agglomeration in the global economy*. Working Paper No. 544, 35 p.

Bellais, R. (2004). *Economie et défense. Nouvelles frontières entre Etats et marchés*, Descartes et Cie, p. 184.

Breheny, M. J. (1988). *Defence expenditure and regional development* (p. 215). London, England: Alexandrine Press Book.

Carrez, G. (2010). *Rapport N°2857 de l'Assemblée Nationale, Annexe N°10, Défense préparation de l'avenir*, 81 p.

Catin, M. (1991). Economies d'agglomération et gains de productivité. *Revue d'Economie Régionale et Urbaine* (5), 565–598.

Congressional Budget Office. (2011). *Long-Term Implications of the 2012 Future Years Defence Program*, June 2011, Congress of the United States, 51 p.

Cour des Comptes. (2004). *Le maintien en condition opérationnelle des matériels des armées*, Rapport – Décembre 2004, 94 p.

Deo, R. B., Starnes, J. H., Jr, & Holzwarth, R. C. (2001). Low-cost composite materials and structures for aircraft applications. Paper presented at the RTO AVT Specialists' Meeting on 'Low Cost Composite Structures', held in Loen, Norway, 7–11 May 2001, and published in RTO-MP-069(II).

Dulait, A., & Carrère, J. L. (2008). *Rapport d'information N°102 du Sénat, Tome VI- Défense-préparation et emploi des forces*, 20 Novembre 2009, 48 p.

Erickson, R. A. (1977, July). Sub-regional impact multipliers: Income spread effects from a major defence installation. *Economic Geography*, *53*(3), 283–294.

Freville, Y. (2005). *Rapport d'information N°426 du Sénat sur le maintien en condition opérationnelle de la flotte*, 23 Juin 2005, 53 p.

Freville, Y. (2008). *Rapport d'information N°352 du Sénat sur la structure intégrée de maintien en condition opérationnelle des matériels aéronautiques du ministère de la défense (SIMMAD) et le maintien en condition opérationnelle des matériels aéronautiques du ministère de la défense*, 21 Mai 2008, 65 p.

Fujita, M., & Thisse, J. F. (1997). Economie géographique, problèmes anciens et nouvelles perspectives. *Annales d'Économie et de Statistique* (45), 38–87.

Fujita, M., & Thisse, J. F. (2003). *Économie des villes et de la localisation*, De Boeck, 560 p.

Gates, E. (2004). The defence firm of the future. *Defence and Peace Economics*, *15*(6), 509–517.

Hansen, N. (1990). Do producer services induce regional economic development? *Journal of Regional Science*, *30*(4), 455–476.

Hartley, K., & Sandler, T. (1995). *Handbook of defense economics* (Vol. 1). New York, NY: Elsevier.

Hartley, K., & Sandler, T. (2003). The future of the defence firm. *Kyklos*, *56*(3), 361–380.

Hartley, K., & Sandler, T. (2007). *Handbook of defense economics. Defense in a globalized world* (Vol. 2). New York, NY: Elsevier.

Hebert, J.-P. (1995). *Production d'armement. Mutation du système français*, Les études de la documentation française, Paris, 1995, 222 p.

Jacobs (1969). *The economy of cities*. New York, NY: Vintage.

Krugman, P. (1998). Space: The final frontier. *Journal of Economic Perspectives, 12*(2), 161–174. Spring 1998.

Leontief, W., Alison, M., Karen, P., David, S., & Edward, T. (1965, August). The economic impact – industrial and regional – of an arms cut. *The Review of Economics and Statistics, 47*(3), 217–241.

Marini, P. (2004). *Rapport général N°74 du Sénat sur le projet de loi de finances pour 2005*, Tome III, Les moyens des services et les dispositions spéciales (Deuxième partie de la loi de finances), 71 p.

Marty, A., Sordi, M., & Viollet, J. C. (2008). *Rapport par la commission de la défense nationale et des forces armées sur l'aéromobilité*, Rapport N°666 de l'Assemblée Nationale, 59 p.

Menderhausen, H. (1958). Economic problems in air force logistics. *The American Economic Review, 48*(4), 632–648.

Metcalfe, L. (1984). The impact of the efficiency strategy. *Public Administration, 62*(4), 439–454.

Meyer, G. (2002). *Rapport d'information sur l'entretien des matériels des armées*, Rapport N° 328 de l'Assemblée Nationale, 56 p.

Paloyo, A. R., Vance, C., & Vorell, M. (2010). The regional economic effects of military base realignments and closures in Germany. *Defence and Peace Economics, 21*(5–6), 567–579.

Panzard, J. C., & Willig, R. D. (1977). Economies of scale in multi-output production. *The Quarterly Journal of Economics, 91*(3), 351–518.

Pyles, R. (1999). *Aging aircraft: Implications for Programmed Depot Maintenance and Engine-Support Costs*, Project Air Force, RAND – Testimony (CT-149), February 1999, 9 p.

Rioux, J. J. M., & Schofield, J. A. (1990). Economic impact of a military base on its surrounding economy: The case of CFB Esquimalt Victoria British Columbia. *Canadian Journal of Regional Science, 13*(1), 47–61.

Smith, A. (1776). Recherche sur la nature et les causes de la richesse des nations (Livres I et II). *Economica*, p. 320

Van Roo, B. D., Carrillo, M., Drew, J. G., Lang, T., Maletic, A. L., Massey, H. G., Masters, J. M., ... Tripp, R. S. (2011). *Analysis of the Air Force Logistics Enterprise Evaluation of Global Repair Network Options for Supporting the C-130*, RAND Corporation Report, 96 p.

Vincon, S. (2002). *Rapport N°350 du Sénat au nom de la commission des Affaires étrangères, de la défense et des forces armées (1) sur la situation et les perspectives des forces aéromobiles de l'armée de Terre*, 41 p.

Weston, D., & Gummet, P. (1987). The economic impact of military R&D. Hypotheses, evidences and verification. *Defence Analysis, 3*(1), 63–76.

FOR A GENERAL CONCEPT OF ECONOMIC AND HUMAN SECURITY

Jacques Fontanel and Bénédicte Corvaisier-Drouart

ABSTRACT

International security is a constant threat to the pursuit of economic optimum. In the traditional economic analysis history, states are seen as agents in constant search for power, which leads to the emergence of conflicts of interests. The modern concept of security can be defined as the economic study of all the risks of short, medium and long term on the functioning of economic and social life. It can be divided into four sublevels: individual security, national security, international security and global security. The adoption of an enlarged approach to international security by integrating economic and environmental conditions highlights the expression of new collective priorities. Today, theories of security take into account the economic, human and social relationships, societal priorities and the balance of power in the international system. Human security implies a multi-disciplinary analysis, including human rights, state organisation, international relations and strategic studies. Security and sustainable development are deeply interconnected, which involves bearable production conditions for the environment in the long term,

The Evolving Boundaries of Defence: An Assessment of Recent Shifts in Defence Activities
Contributions to Conflict Management, Peace Economics and Development, Volume 23, 75–96

ISSN: 1572-8323/doi:10.1108/S1572-832320140000023007

the end of extreme poverty, the creation of social stability and the rejection of discrimination.

All definitions of security depend on points of view, specific predispositions and basic values. The economics of security can be defined as the economic study of all the risks of short, medium and long term on the functioning of economic and social life. It concerns information, prevention and management of insecurity in the economic and political systems, affecting public and private, legal or illegal sectors. However the concept of 'risk' is polysemous. It describes both the probability that a desired (or undesired) event occurs and the volatility of economic factors due to exogenous shocks, speculative phenomena, adverse effects or random events. The bottom line of security is survival. The modern concept of security can be divided into four sublevels: individual security, national security, international security and global security.

However, from the beginning of the history of economic analysis with the mercantilist school,[1] states are seen as agents in constant search for power, which almost always leads to the emergence of conflicts of interest. Political instability derives from economic instability. They argued that the successful state is that state which exports more than imports, because gold reserves currency was an assurance for peace, power and stability for the country. Economic and individual interests are subordinated to the pursuit of state power. Military force is then considered as an instrument of national economic development,[2] since it is supposed to provide sufficient security for economic and social actors to innovate and carry out their functions of production, consumption and trade. However, the security dilemma is immutable, protecting themselves by arms; the state becomes a threat to its counterparts, except in case of reliable alliances.

According to the liberal thought, state is subordinated to the market and economic agents' interests. It may intervene in the national economy only when some factors jeopardise the stability of the market. Then states have to play a marginal role in the international community and for international economic security, which mainly concerns markets and economic actors. Free markets and change are guarantees for economic and political stability, and then for security.

Nevertheless, Lenin underlines that liberalism is exploitative, competitive and violent. According to the Marxists, proletariat (the base of the system) is not allowed to have more economic security than decided by the capitalists that own production factors. The only possibility for enforcing the labour class' economic security is the revolt against the system and the interruption of this vicious circle. However one may note that, opposed to political violence, French utopian economists of the nineteenth century contested capitalism and anticipated a peaceful evolution towards a new social system, often named socialism.[3]

Then, the historical understanding of economic security combined three levels: macro level (mercantilists) where only military and economic security of the state mattered; micro level, where markets and free traders are the core centre of economic security; and philosophical level (Marxists), where economic security is mainly realised with the end of the exploitation of labour and Proletariat.

During the Cold War, the concept of national security has often been replaced by that of international security, which includes the situation of mutual hostage-taking incident of the great powers in the strategy of nuclear terror. The possibility of a nuclear war, including weapons of mass destruction, was the main threat to international security. However, after the oil crisis, the waves of financial crises started and the international community realised that being safe without stable economy and economic order was unrealistic.

After 1990, military, economic, social, environmental and political threats became part of the complex understanding of security. A sharp increase in capital flows to developing countries in the early 1990s reinforced positive views of globalisation, but that perspective did not last as successive financial crisis affected Mexico, Asia, Russia, Argentina, then dollar and euro. The concept of economic and human security is then related to values that are not neutral. It includes various categories such as national security (conventional analysis), intrastate security (civil or ethic war or conflicts), enlarged security (environmental and economic organisation) and people security, with the survival issue of societies, groups and individuals. The school of thought 'Freedom from Fear' considers that human security has mainly to protect individuals from violent conflicts. Inequalities, poverty or pollutions are not to be fixed this way. Limiting the aim of human security to the end of violence fight (by conflict prevention, emergency assistance or citizen protection) seems a more manageable system. For the school 'Freedom from Want', a more holistic thought is necessary. Hunger, natural disasters and inequalities kill far more people than all international, national or ethnic wars. More generally, human security describes the condition of existence of a society, with the respect, for every people, of the non-violence behaviour, the satisfaction of basic material needs and the respect of human rights.

This new perception has gradually promoted the adoption of an enlarged approach to international security by integrating economic and environmental conditions, then highlighting the expression of new collective priorities.[4] Today theories of security take into account economic, human and social relationships, societal priorities and the balance of power in the international system.[5] Human security includes general vulnerabilities of all individuals and not only of the state. People-centred view of security is essential for the state, regional and national equilibriums. Human security implies a multi-disciplinary analysis, including human rights, state organisation, international relations or strategic studies.[6] It will be interesting to analyse first international security, then economic security and, in a broader perspective, human security in relationship with the two former sectors of people security.

THE INTERNATIONAL SECURITY IN QUESTION

Safety first expresses the subjective perception of a threshold of actual, perceived or feared vulnerability that amplifies or reduces the economic (and therefore social) potentiality for peace and economic development. It is also a qualitative aggregate form of risk relating both to natural forces (i.e. tsunamis) and destabilising technological developments (such as information control), economic (i.e. the effects of destabilising international speculation), social (such as uncontrolled international migration), environmental (i.e. global warming), strategic (such as continuing armed conflicts in the world) and political (including terrorism) areas.[7] There is no optimal level of safety in itself, because 'the economics and politics each have distinct approaches with different dynamics operating at different speeds'.[8]

Because the security of a country is not included in a reliable model, economic policies based on optimum calculus are in fact impossible to build without very strict hypothesis that are not confirmed by the observation. If economists have analysed for centuries the issue of war and security, global economic analyses on these issues are not numerous and riddled with general assumptions often far from the global strategic realities. The aim of national economic security is to protect the economic development of a country that is confronted with multiple attacks of various types, including financial fraud, cybercrime and cyberwar, strategic addictions or fight against industrial and research espionages. Economic security of the territory becomes essential in the context of an open economy full of opportunities but also dangerous with the potential situation of predation of national strategic assets in key sectors.[9]

If the main role of the state is to protect all citizens from the danger of real or imagined enemies, the government has to give top priority to national defence. In this context, military power will is a permanent task for government leaders to ensure national defence service or to seek international hegemony or power. Machiavelli wrote in the early sixteenth century that laws and weapons were the priority of the foundations of the state. However there is a constant contrast between structured states and the invertebrate international order. International laws are not restrictive enough to change substantially the behaviour of international actors such as the great powers. The international system is anarchic because collective rules, when they exist, are inadequately respected, particularly due to the lack of a strong centralised government.[10]

Several quantifiable or observable facts may be highlighted to demonstrate the quality of international security.

The number of conflicts gives interesting data on the intensity and spread of international insecurity. In 2011, 30 conflicts are listed, with several major generic conflicts in Afghanistan, Iraq and Arab countries. The actual situation is more worrying than the situation of 1990 that put an end to the antagonism between capitalist and socialist systems.

The number of nuclear warheads also shows a significant dangerous international security situation. In 1986, there were 70,000 nuclear warheads. In 2011, as a result of agreements (SALT I, SALT II, START I and START II), they are just under 20,000 (Table 1). However, U.S. forces

Table 1. Stocks of Nuclear Weapons in the World from 1960 to 2010.

Years	USA	Soviet Union/Russia	United Kingdom	France	China	Total
1960	20,434	1,605	30	–	–	22,647
1970	26,662	11,643	280	36	75	38,696
1980	24,304	30,062	350	250	280	55,246
1986	24,401	45,000	300	355	425	70,481
1990	21,004	37,000	300	505	430	59,239
2000	10,577	21,000	185	470	400	32,632
2006	10,104	16,000	200	350	200	26,854
2010	10,000	7,200	190	300	280	17,970

have an explosive power of 2,400 megatons, or 100,000 times that of Hiroshima and a 'bang for buck' very high.

In 2011, nuclear forces are still very important in the strategies of states, but their proliferation is a potential danger for the world security. The interest of partial disarmament of nuclear weapons for peace must be reduced by the danger of their proliferation,[11] technological advances and the loss of the important role of the Conference on Disarmament.[12]

The proliferation of conventional weapons is picking up again. There are over 700 million firearms in the world that feed civil wars and the UN control endeavours of such weapons have not yet been successful.

The indicator of military spending is not a good indicator of the level of international security. After the Cold War, military spending declined, because of the difficulty to conceive and organise new military strategies, the public finance crisis and the collapse of economies in transition.[13] The excessive burden of military spending relative to GDP of the USSR is probably a factor in its economic exhaustion.[14] Military spending in excess becomes a burden on national economies. However, today, governments saw positive gains to their national economic and technological base, and ultimately to their military power, through links to the global economy. Changes in military technology had opened an era of spin-on from the civilian economy rather than spin-off from the military sector.[15] In this context, states that rejected expanded international economic exchange take the risk to develop a conventional military inferiority.

The international military, financial or political alliances such as NATO,[16] International Monetary Fund (IMF) and UN aim to reduce uncertainty by establishing international standards improving the international system cohesion. UN urges dangerous but useful peacekeeping operations, when the parties involved are willing or forced to find a peaceful solution.[17]

Community security requires that states that make up the entire political world reject the use or threat of force as a mechanism to solve their mutual conflicts. Thibaud Richard[18] considers that democracy, as a system, is not particularly peaceful except with regards to the respect of his fellows.

The search for the cost reduction of security should be considered insecure. However, it is not clear that deterrence policies aiming to increase the marginal cost of terrorist attacks are effective against the typically low costs involved in such transactions (the attack of 11 September 2001 cost less than 500,000 euros all included). Alternative policies seek to reduce the benefits of terrorism. States must choose between various policy instruments, from information, dedicated institutions, regulatory coordination or penalty deterrence. The private sector, much more preoccupied with short-term issues due to the economic competition, needs a state financial aid for the protection of their activities against international illegal attacks.

State security policies can have pernicious effects. A tightening of border controls brings delays, increases the shipping time and indirectly encourages the profitability of illegal operations. Provisions made in the field of national defence, fight against crime and civil rights impose costs on businesses. In total, the damage of terrorism worldwide was estimated at $75 billion per year, on the short term.[19] The Organisation for Economic Co-operation and Development (OECD) estimates that, apart from the tragic loss of lives, the economic cost of the terrorist attacks of 11 September 2001 reached $120 billion, but the value of this estimation must be questioned.[20] In order to manage insecurity, states are developing new agencies, new rules or laws, affecting the civilian management of natural disasters, industrial accidents, infectious diseases and international terrorism.

Security research can use the functions of ownership (intelligence), ban (restricted access to information) and manipulation (intoxication). Sticks of dynamite, gasoline, fuel oil or fertilizers create explosions that can damage the nodes of information networks, causing a threat to the evolution of large financial sums that pass through them. However, these procedures are difficult to detect and are accessible to the greatest number of terrorists. Andres Behring Breivik, with his action in Utoya Island and government building in Oslo, gave a terrible confirmation of the permanent world insecurity and the huge cost for its prevention.

Moreover, computer viruses or infectious are more global and destructive threats. Blocking the remotely commanded centres computers or information transmission is a powerful weapon in the hands of an enemy that can disrupt the functioning of a country's system of defence. There are several methods of attack, from propaganda disinformation, data collection, to confidential sabotage of infrastructures and systems (pipelines, transport, nuclear power plants, etc.). The destruction of sensitive equipment to intercept orders command is a goal against an enemy. The electronic attack prevents the opponent from using their electronic spectrum. The cyberattack can also be used to neutralise or destroy a country's military infrastructures in its C4ISR ability (command, control, communications, computing, intelligence, surveillance and reconnaissance). The Chinese army is also tackling the sensitive civil and military technologies.[21] Russia used such weapons against the Baltic States and Georgia in the South Ossetia conflict. States must collect information, conduct research on defensive technologies and share them with industry when necessary, for educational purposes and defence policy.

The concept of security cannot be reduced obviously to the active and passive defence areas; it relies on the specific economic, environmental, societal and political context. It has to take into account issues relating to pollution, terrorism, ethnic and regional conflicts, famine, drugs and crime, as well as religious fundamentalism. The collective interest of humanity needs economic security for all states and individuals. With the use of economic weapons,[22] the importance of the defence industrial base, the role of natural resources or the emergence of the economic integration process, the national search for security becomes more complex. Globalisation transforms the definition of economic security, with a growing weight on non-military factors.[23] It supposed positive benefits were not sufficiently proved for the well-being of all nations, and then for every citizen.

The feeling of insecurity is growing with the importance of unemployment, public and private debts, inequalities, social system crisis or the lack of job security. Human security, which had considerably improved during the last century, is beginning to decline in central cities. Terrorism tends to exploit the critical dependency of modern societies on energy, transport, financial services and information systems. Protecting these sensitive areas is necessary for the world community.

INTERNATIONAL INSECURITY, A CONSTANT THREAT TO THE PURSUIT OF ECONOMIC OPTIMUM

Strictly speaking, the concept of economic security is analysed on the basis of the importance of existing conflicts, changes in military spending indicators and domestic spending for the intelligence and domestic security (including the fight against terrorism). However, in a larger sense, economic security includes development issues, such as hunger or poverty. Industrialised economies experience new anxieties. How to explain to developing countries that the kind of economic development that developed countries have enjoyed is no longer applicable to them? They have to realise that the affluence of raw material and energy that developed countries have lived with was possible because of international social inequality and are now challenged by the development of emerging countries. It is no more a question of three or four decades in a positive-sum game, but at the best in a zero-sum game. Access to raw materials and energy resources as well as free competition of labour and capital create conditions of uncertainty that the powerful countries are willing to contain. Ultimately, the use of military power can be a claimed remedy.

Insecurity is part of modern societies' functioning based on competition and power relations.[24] It is therefore necessary to prevent, mitigate or eliminate the various risks that may affect people's lives or interests. War, terrorism in all its forms, crime, pollution, ecological threats, possible disruption of global supply chains, particularly energy, espionage or computer viruses justify the world overall expenditures of $2 trillion in 2011.

The structural analysis of economic security highlights six levels affected by economic security and reacting differently to threats: individual human beings, enterprises, states, interstate civil and military relations, transnational and global relations. The market is not in itself a factor of stability and economic security. For B. Udovic, three divisions come up in the analysis of economic insecurity: macro, medium level and micro factors (Table 2) (Udovic, 2011).

Table 2. Divisions of Economic Security Factors.

Macro Factors	Medium-Level Factors	Micro Factors
State, interstate relations, world	Enterprises, firms, business associations	Individual persons, families, societies
Stability of the state markets, Free trade, Competition, Sustainable development, GDP growth, Productivity, Low inflation rates, Low unemployment rates, Stable exchange rates, Balance of payments equilibrium, Indebtedness, Stabile 'endowment' with production factors (such as oil and gas), Avoiding and reacting to speculative attacks, Solving problems connected with drugs using, trafficking and criminal groups.	Stability of state macroeconomic environment, Innovation, new inventions, Marketing, Solvency and financial discipline, Flexibility, Stabile 'endowment' with production factors, Technology diffusion, Flexibility of administration, Stable exchange rates, Lean production, Ethical dilemmas, Knowledge, Minimising black market production.	Stability of state macroeconomic environment Food, water, shelter stability, Lodging, Job security, stability, Stabile and 'fair' wages, Trust in institutions, Minimising (absolute and relative) poverty, Minimising social exclusion (maximise individual reputation), Education, Minimising phobias and various –isms, Developing free movement of individuals, Avoiding the 'magic circle' of living standards and employment.

The issue is to understand how to reduce the uncertainty among economic agents and to improve their performance. Markets are likely to appreciate government's efforts to secure production and trade.[25] However, in the context of globalisation, it is unclear whether the effects of these expenditures are equal for a country to the opportunities to invest. Countries that do not respect international safety standards cannot benefit from globalisation; they will face high premiums risk, with a reduction in the contribution of foreign investment.

The search for security involves both benefits and costs, including transaction. There is a choice between adequate security and economic and social efficiency. An analysis from the World Bank concluded in 2003 that trade liberalisation, harmonisation of customs and international coordination would increase the international exchange of 75 countries up to $377 billion.[26] However interdependence and openness can increase the reasons and general feeling of instability. The question is always the same. To which extent interdependence and openness are possible bringing costs in terms of insecurity lower than the benefits free trade allows?

Coalition building security should be engaged in a global economy to reduce its costs and improve negotiation processes. Exchange of useful information and joint actions have been carried out after 11 September 2001, events even if the initial proposed measures, particularly with regard to tax havens (to fight the terrorists forces), were not ultimately implemented. Today, out of the military, infrastructure, security automation, surveillance and information sharing in the ports, airports and border posts have been highly developed. Access to services and security products raises the crucial issue of equity. It is up to governments to ensure access and equitable participation in national economic security.

Increasing liberalisation of the economy (goods, capital, financial assets, and human beings) is a weakness in terms of differentiating national rights. The finally asymmetric weight of hegemonic nations on international regulations constitutes a problem for their interpretation. Globalisation produces some pernicious effects, as new technologies both facilitate criminal acts and provide substantial financial compensation. New technologies in identifying and monitoring are being developed in both the private and public sectors, following the example of biometrics or satellite surveillance. There is no doubt that the market for civil security will tend to grow rapidly, around 7–8% per year according to the OECD.[27] The risks are mainly related to the scale of terrorist violence, global energy power failures, the rise of computer viruses, frauds involving electronic purchases, strikes or embargoes affecting the distribution of oil and gas and international financial instability. Political leaders have to define a range of policy instruments to reduce, manage or compensate insecurity. However, 'most studies on the impact of economic security are based on assumptions which we find strong evidence in the conclusions'.[28]

Globalisation has produced new situations of economic security centred on five main types of unwanted transmissions across national borders:

The economy is a war weapon.[29] States may use this means to compel a foreign country to meet its own goals, thus reducing international security. A strong economy is a safety factor. Under these conditions, states seek to strengthen their economies by all means: hidden subsidies, weapon shortages, embargo, boycott, frozen foreign assets or financial use of the weapon threats. 'The relentless pressure of global competition threatens the solidarity, the invisible heart of human development ... We must rethink national and global governance, by focusing on human development and equity'.[30] There is wealth drainage from South to North and a standardisation of culture. So far sustainable development is a theoretical concept that globalisation gums in all daily activities.

Illicit financial flows are more difficult to check and easier to disguise as legal economic transactions (terrorism, crime, pollution). Globalisation offers a powerful breeding ground for the development of transnational crime. With regard to organised crime, current annual revenues of illegal criminal activities are gigantic. Domestic violence is also concerned. Multinational crime is settling over with the Sicilian Mafia, Chinese triads, the Japanese Yakuza, the Colombian cartels of Medellin and Cali. The crime has an estimated burden of 20% of U.S. GDP and 7% of the British GDP, taking into account all the costs – from prevention to prison expenses, to tangible and intangible costs, physical and psychological damages.

The potential development of terrorism is important. In the United States, the private sector for homeland security, which has been developed following the 9/11 events, has an estimated annual turnover of more than $10 billion, after a 'peak' in 2003 evaluated between $46 and $76 billion. The public sector (federal, local governments) now spends more than $30 billion per year. For the OECD, these measures lead to an increase in the annual cost of trading around 1% or $75 billion.

New scarcities may result in armed conflicts. Water is a key strategic issue in regions facing scarcity. Long regarded as inexhaustible for centuries in developed countries, fresh water becomes an unstable, cyclical, seasonal, vulnerable and regionally rare resource, whose consumption is necessary to human survival. The Middle East is a particularly sensitive region in terms of scarcity and misallocation of water resources. Today, it faces high population growth in a dry climate, with inequalities in provision of resources and levels of consumption, dependence of neighbours sharing the same pond, complex river systems and weakness of the collective management. It is a new source of insecurity.

At last, international economic shocks, from the last financial crisis, can undermine economic growth, increase inequality, raise economic vulnerability and lead to new international and national military tensions turning into wars or revolutions. Neighbours or economic partners might manipulate national economic vulnerability. Today governments have a developing consciousness on dangerous spill-over effects from excessive globalisation process and the risks of economic complexity allowed by economic openness. National policies remain central, although those policies and the institutions that produce them will be in a continuous adaptation. However, they are more and more dependent on regional institutions that offer complementary help to more stability and less uncontrolled conflict situations.

In general, private spending on security includes security consumption and investment for households and businesses (alarm systems, safes, security systems), but also insurance payments. It mainly concerns the developed world in general, first of all the United States (which represents 40% of global private expenditures, or more than $100 billion, excluding spending on national defence). There are both a substitution effect (which changes actions and security instruments to reduce the perceived risk) and an income effect (which leads to a slowdown in economic activity generated by the new risk). Insecurity imposes three types of costs to individuals or communities who express an aversion to risk: costs based on the event itself (theft, fraud, property damage, health, war or destruction), direct costs related to the response agents found to these threats and indirect costs caused by the secondary action of a government in relation to the reactions of households and businesses.

The market and state may require security measures (in the areas of environment or labour, for example) that everyone is compelled to respect according to its condition and action. The raising perception of insecurity leads to increased transaction costs related to the exercise of a business, including transportation costs. Thus, the piracy off Somalia brings a social issue on tuna fishermen, cargo traffic or leisure tourism that use this space to thrive. A decline in trade may reduce economic activity and enhance geographical partitioning. The state may intervene to maintain the advantages of a national economic growth, avoid confrontation and develop social foundations for an economic development that an insecure situation would slow down. A priori urbanisation can produce a better collective defence, but cities are also becoming target areas for terrorists, particularly in transportation.

In peacetime, the quest for hegemony over information networks is both secret and essential. In this matter, oceans barriers or military weapons do not improve international security and the control of security for communication tools. The spreading of information in the world is faster, cheaper and simpler and its resonance is difficult to counter. The interconnection of computers reduces the security of both states and their inhabitants. Vulnerabilities are more important and powerful states influence the evolution of intangible capital by imposing standards and international law in favour of national goals, setting up surveillance systems for targeting the global exchange of information (such as the Echelon system) and monitoring networks. The threat of using military power becomes legitimate. The Pentagon now has a Cyber Command, with 8,000 people, responsible for responding to computer attacks and offensive conducted in the cyberspace.

In addition, the global polarisation of wealth threatens peace and security in the coming years. In other words, some relatively poor countries will have access to nuclear weapons or cyber-weapons to claim a larger role in the distribution of wealth. In this context, war or threat of war may grow in the future. The question of whether 'hungry' states are able to rise up has always been debated. The majority of conflicts that are not part of U.S. policy and U.N. policeman of the world are national by nature, in a rejection of the state monopoly of violence. Today's 'Warlords', mercenaries and 'off-the-law' are seeking to take control of scarce resources, with as greater ease as the small arms trade is flourishing. Great powers 'strategists' now insist on the protection of developed countries' interests against the growing threats in the South in terms of terrorism, civil war, production of mass destruction weapons (including nuclear) and international conflicts aroused by 'rogue states'. The political insecurity and economic crisis increase with the development of nationalism and religious fundamentalism, the collapse of state management, illegal migration, ethnic conflicts and economic underdevelopment.[31]

The state action, mainly the economic intelligence, is therefore essential and will have to grow to protect the country's economic and technological heritage.[32] Two types of risks are concerned: the direct dependence in strategic sectors and the management of knowledge services that hold, exchange and process sensitive information. The information warfare consists in a set of methods and actions to put a rival in trouble or become involved in the acquisition of information (data or knowledge) to degrade the power of the opponent. Cyber warfare must be prepared, at least for national defence. The expansion of ICT infrastructure (including broadband infrastructure) appears to contribute to the vulnerability of information systems.

The blocking of remotely command centres, computers or information transmission is a powerful weapon in the hands of a competitor or an enemy that can disrupt a country's system of defence. In France, the Central Directorate for National Intelligence is designed to manage the intelligence service, fight against terrorism, intelligence management as well as social movements and corporate events' analysis. An 'against-interference' service has been created, in charge of fighting industrial espionage. These industrial spies aim at tracking the most sensitive technologies to protect French companies from being looted, not just informing them of the threats but by hiring and training specialist in the struggle against espionage industry. The new Public Procurement Code provides national security in the choice of suppliers.

ECONOMIC AND HUMANITARIAN CONCEPT OF SECURITY

Human security is a more people-centred approach than a state-centred one, but its definition is still vague. In addition, security has to do with the general issue of freedom, balance of individual rights and respect for privacy. The choice between economic freedom and security is often painful, because high levels of regulation and restriction threaten productivity and the optimisation of utility. However, the networked economy has also to deal with security, because the vulnerability of interdependent and connected data systems has increased. Government agencies' action is needed because the security of a state, national groups and individuals is a 'public good'.

Economy has not only to be referred to economic science. It has a political and philosophical substratum that is currently omitted. There is no doubt that philosophical thought is essential to introduce an analysis on what is efficient for people in an unequal society, and for whom it is. For François Perroux,[33] there are three types of human costs: feed people, provide health care and free the slaves. Let us add an essential criterion: the general physical, economic and humanitarian security.

For United Nations Development Programme (UNDP), globalisation is not a new concept, but today it has been characterised by the narrowing of both distances and time and by the removing of all barriers, from trade tariffs to financial assets and labour. If markets are a guarantee of efficiency, they do not stand much for equity neither for the expression of public goods (such as the environment). For UNDP, one must defend ethics (fight against violations of human rights), equity (reducing national and international disparities), integration (rejection of exclusion and marginality), human security (reducing the system instability and improve the respect of collective rules), sustainability (including the conservation of environment) and development (decline in poverty). However, the values set by UNDP are not always met by the current globalisation, a project of universal reason that exacerbates the liberal identity crises and forgets safety. Globalisation, as the final step of the liberal world, can lead to everyday violence. Indeed, there are still more than 1.3 billion people living in unworthy poverty.

World consumption of raw materials has increased by 2000% since 1900. There is a depletion of fish stocks, overgrazing, desertification, chemical pollution; American industry discharges 150 times more chemical waste than it did in 1950: the risk of developing cancer is twice as big for an American as it was for his grandfather.[34] Companies assume that profitable economic activity and job remuneration suppose irreversible environmental damages. Scientists predict an increase of 2–10 degrees in average global temperatures in the coming century, ceteris paribus. Under these conditions, the environment is getting more hostile with cyclones, dust storms, forest loss or lack of water. Normative in nature, collective security research has a high subjective value. It depends of the political and societal goals, in competition with social actions. When human needs are not a priority, states refer to the idea of necessary but temporary collective sacrifice for a better future. One has thus to choose and make priorities between present and future, between the repair capacity allowed by anticipated new technologies potentialities and the irreversibility of destructions or pollutions. It is a risky gamble on the future.

International security has to be based on the concept of sustainable human development, which involves bearable production conditions for the environment in the long term, the end of extreme poverty, the creation of social stability and the rejection of discrimination.[35] The will of growth at any cost has led to the development of unacceptable inequalities, seeds of conflict. Economic growth is today characterised by a forced process of globalisation, leading to internal and external crises that increase the relative poverty in a global system where priority is given to dominant forces' growth, as an illustration of human progress. Insecurity lays on the opening of the world. The internationalisation of production systems has led to a fragmentation in specialised business functions (R&D, technology development, production, distribution and marketing). The growing geographical dispersion of productive activities organised in a global supply chain makes disturbances, delays and frauds more frequent and less punishable.

A Human Security Index has been released since 2008. It takes into account some Human Development Index items (education, health, and income) as well as an equitability indicator that changes the results. A Social Fabric Index analyses human security with respect to environment, diversity, peacefulness, freedom from corruption, and information empowerment. Both have been mixed in a prototypal index called Human Security Index. The perception of security is based on a social structure, as the result of a subjective consensus able to make the governments' actions change over the time, but it is an imperfect indicator of the international and national situation. Societal security highlights cultural, religious, linguistic and national identity issues. It raises the question of traditional communities' survival. The measurement of Human Security is difficult and suffers from the lack of consensus in data definition and interpretation. However, there is no doubt that it is an interesting instrument to understand the international security situation.

Based on the Global Peace Index that measures their ability to maintain and promote peace, 144 countries data were analysed. As usual, New Zealand, Denmark, Norway, Iceland, Austria, Sweden, Japan, Canada, Finland and Slovenia are defined as the most peaceful countries. These surveys do not take into account the protection these countries benefit directly or indirectly from military alliances, allowing them to be within the 'free riders' of global security.

Security and development are deeply interconnected. Human security and human development consider that people welfare is the ultimate goal for them. Thus, a multidimensional analysis is fruitful. First, human security is a central aim for economic development, because it enlarges human choices as well as society and individual's quality of life. Second, lack of internal and international security obviously has a negative impact on economic growth. Markets need security of economic exchanges, a justice to judge the commercial conflicts and a police to stop illegal activities and violence. Finally, unbalanced development boosts predation trends. Economic security requires a human response to chronic threats such as famine or disease. Famine can occur without anyone's rights being violated. For Amartya Sen,[36] one should give as much attention on the unequal distribution of substantive freedoms, entitlements, and capabilities as on income. Similarly, for economic security, both inequality and poverty are the roots of individual vulnerability.

Human security is the primary purpose of organising a state. Economic security involves welfare for individuals, with an assured basic income from a remunerative job or from a public safety income. Today, more than 50% of people in the world live with the feeling of precariousness. Security requires that people have economic access to basic food. There is a large world production of food, but inequalities and the lack of purchasing power are the main factors of famine. For the UN, access to work, assured income, disease protection, admittance to health security, malnutrition, sufficient level of education, clean water and air resources, respect of traditional or ethnic peaceful relationships, the respect of human rights and personal protection against violence are the main dimensions of economic security policy. However, poverty is usual today in a lot of emerging and developed countries, and famine still exists in developing countries. Without a control of inequalities, classes' tensions, ethnic violence and regional conflicts bring about a general insecurity.

Mainly, human security is clearly connected to the level of economic and social development. However, globalisation is not as efficient as needed for national economic development. When only few emerging countries seem to reach the first step of development, poverty is growing in developed and developing countries. The dominant paradigm of liberal economy ignores the debate on inequalities, considering it as irrelevant in the short term. It considers that economic development implies social sacrifices in areas such as health, education, the environment and employment. However, human development and human security are inextricably linked since progress in one dimension enhances the chances of progress in the other. Well-being strengthens stability and human security sustainability.

State sovereignty and non-intervention principles did not justify international intervention in national disputes. However, with the human security concept, the world order's sake can legitimate a state or an international organisation in some circumstances to get involved in another state's internal disputes. In 2001, the International Commission on Intervention and State Sovereignty determined how the 'right of humanitarian intervention' could be exercised. The protection of individual welfare is more important than the state and a priority should be given to prevention. Humanitarian intervention is a contentious issue.

CONCLUSION

Traditional security concerns the state capacity to defend itself against external threats. All the interests are subordinated to the state, which has to protect national boundaries, values, institutions and population. Thus, national defence refers to the ability of the state to deter or defeat an attack to protect its territory, maintain its integrity or build up national power at the international level (in an idea of hegemony, international or regional military power). State is the only actor centralizing public decisions and deciding of the security strategy. The United States' current actions seem to develop the questionable effects of 'American imperialism' with a 'clash of religious civilisations' in return.

For Todd Sandler,[37] prevention and protection have different externalities. A safety equipment aiming at reducing the probability of a collectively undesirable phenomenon can be invested in at the expense of other targets (negative externality), which often leads to insufficient investment as far as security is concerned. Thus, the protection of a site against terrorism leads to increased security costs, which, ceteris paribus, will not be used on another site, increasing the overall risk. Finally, compensation schemes alter the agents' reaction to the risk and tempt them to play on moral hazard.

A growing interest for both a redefinition and new analysis of economic security comes up from the recent societies' troubles in the globalisation process. Today, military action is to be reconsidered, moving from the traditional predation and hegemony purpose to the defence of economic interests and research of alternative strategies to terrorise terrorism.

A people-focused human security, mostly aiming at protecting the well-being of human beings, must replace the traditional concept of security and consists in the state's protection from international aggression, environmental threats, infectious diseases, economic recession, poverty and inequality. The involvement of regional and international organisations, local communities and non-governmental organisations is needed. Citizens of all countries have to contribute to the human security by claiming loud and clear for solutions to insecurity.

However, the rise of the national security paradigm after the 11 September 2001 events demonstrates how traditional security is still prevailing. Military expenditures are 20 times more important than the expenses for international aid, including in the countries were UN or the United States carry out peacekeeping operations. Human security has to be implemented, thanks to domestic policies on development and human rights. The main question is the prioritisation or 'hierarchy' of the threats on security. The famous dilemma 'food or guns' is then more largely revived.

What are the main situations of insecurity, for whom and where? An international agenda must be settled in the frame of a negotiation process involving international organisations, states, non-governmental organisations, private companies, financial institutions and some leading scientists[38] (for a comprehensive and more neutral comprehension of the origins, importance and solutions for all conflicts in the world). Interventions must be targeted, implemented, monitored, and coordinated.

In militarised societies, the reallocation of military resources requires political, social, ideological and economic adjustments. A strong political has thus to be developed with the support of public opinion. International financial institutions like the IMF, the World Bank or regional development banks have a significant influence on the poorest states' economic policies, but they barely take into account conflict prevention in their mandates.[39] Today, they mostly consider the importance of military spending and are less interested by structural adjustment programs that have contributed to the development of economic inequality.

Security is a fluctuating concept, but it concerns directly world economy, which provides the means for the armed security to implement, create or eliminate the conditions of international conflict and offer more 'capabilities' and entitlement to all people. It is impossible for an economist to determine the optimal level of international security, knowing how complex is the task of quantifying the objective values and subjective preferences, costs and benefits of security, private gains and social gains, direct and indirect impacts.

However, economic interdependence could provide security benefits. For instance, when some governments try to reduce perceived vulnerability, their action has a positive effect on the global energy markets. National institutions do not provide their own economic security on their own: regional and global institutions can also provide alliances, insurance, solidarity, adaptation and credibility. Regional and global institutions sometimes offer interesting complements to national security.

Insecurity is part of modern societies, which philosophy is based on competition and power relations. Competitive capitalism is mostly built on permanent insecurity for all economic actors. Then, economic security seems to be incompatible with a pure market-economy system, at least in the short and medium term.

NOTES

1. Fontanel, Hébert, and Samson (2008).
2. Smith and Fontanel (2008).
3. Fontanel, Bensahel, Coissard, and Echinard (2008).
4. Sheehan (2005).

5. Fontanel (2005b) and Fisher (1993).
6. United Nations Development Programme (1994).
7. OECD (2003, 2004).
8. Smith (2010).
9. Fontanel and Coulomb (2008).
10. Fontanel and Chatterji (2008).
11. Despite the UN Treaty on the non-proliferation of nuclear weapons initially signed in 1968 and extended in definitively in 1995, there are, respectively, 180, 60, 40 and 10 nuclear weapons available in Israel, India, Pakistan and North Korea.
12. Natural Resources Defence Council (2006).
13. Fontanel (1995).
14. Fontanel, Coulomb, and Samson (2001).
15. SIPRI yearbook (2009).
16. Sheehan (2010).
17. Sheehan (2003).
18. Richard (2009).
19. Walkenhorst and Dihel (2004).
20. Fontanel (2005a).
21. Rutherford (2009).
22. Fontanel and Bensahel (2002).
23. Fontanel and Antipas (2004).
24. Buzan, Waever, and de Wilde (1998).
25. Brück (2004a).
26. World Bank (2003).
27. Stevens (2004).
28. Brück (2004) and Brück and Wickström (2004).
29. Kauffmann (2006).
30. PNUD (1999).
31. Sen (2002).
32. Such as in France, with the circular on the economic defence published in the *Journal Officiel* on 23 March 2002.
33. Perroux (1952).
34. Transport facilities drive to ecological nonsense. According to the Institute Wuppertal, the components of strawberry yoghurt crossed more than 1,000 kilometres before being mixed, what increases wrongfully energy consumption. The enterprise interest does not match the long run collective need.
35. Bensahel-Perrin, Fontanel, and Corvaisier-Drouart (2009).
36. Sen (1981, 2002).
37. Sandler (2003).
38. Fontanel (2005b).
39. Kauffmann (2006).

REFERENCES

Bensahel-Perrin, L., Fontanel, J., & Corvaisier-Drouart, B. (2009). *Les organisations non gouvernementales, L'homme au cœur d'une mondialisation solidaire*. Paris: Collection La Librairie des Humanités, L'Harmattan.

Brück, T. (2004). Économie de la sécurité : Arbitrages économiques. In *OECD,* l'économie de la sécurité. Paris: OECD, 115.

Brück, T., & Wickström, B. A. (2004). The economic consequences of terror: Guest editor's introduction. *European Journal of Political Economy, 20*, 293–300.

Buzan, B., Waever, O., & de Wilde, J. (1998). *Security: A new framework for analysis.* London: Rienne Liener.

Fisher, D. (1993). *Nonmilitary aspects of security. A systems approach.* Geneva: United Nation Institute for Disarmament Research, UNIDIR.

Fontanel, J. (1995). *Les dépenses militaires et le désarmement.* Paris: Publisud.

Fontanel, J. (2005a). *Globalisation économique et sécurité internationale. Introduction à la géoéconomie,* Avant-propos de Kenneth Arrow, Prix Nobel d'économie, Collection Côté Cours, Université Pierre-Mendès-France, Grenoble.

Fontanel, J. (2005b). *La globalisation en 'analyse'. Géoéconomie et stratégie des acteurs.* Paris: Collection Librairie des Humanités, L'Harmattan.

Fontanel, J., & Antipas, T. (2004). Cancun et l'Afrique, la guerre du coton. *Géoéconomie africaine,* Paris.

Fontanel, J., & Bensahel, L. (2002). Les stratégies de la guerre économique. In *Globalisation économique et sécurité internationale. Introduction à la géoéconomie,* Collection. Collection Côté Cours, Université Pierre-Mendès-France, Grenoble, pp. 29–50.

Fontanel, J., Bensahel, L., Coissard, S., & Echinard, Y. (2008, October). French utopian economists of the nineteenth century. *Defence and Peace Economics, 19*(5), 339–350.

Fontanel, J., & Chatterji, M. (2008). War, peace and security, Preface Walter Isard, *Contributions to conflict management, peace economics and development* (Vol. 6). London: Emerald.

Fontanel, J., & Coulomb, F. (2008, October). The genesis of economic thought concerning war and peace. *Defence and Peace Economics, 19*(5), 321–330.

Fontanel, J., Coulomb, F., & Samson, I. (2001). Military conversion and transition in Russia. *Pax Economica,* n°6.

Fontanel, J., Hébert, J.-P., & Samson, I. (2008, October). The birth of the political economy or the economy at the heart of politics: Mercantilism. *Defence and Peace Economics, 9*(5), 331–338.

Institute for Economics and Peace. (2013). Measuring the state of global peace. *Vision of Humanity.* Retrieved from http://www.visionofhumanity.org/pdf/gpi/2013_Global_Peace_Index_Report.pdf

Kauffmann, M. (2006). *Gouvernance économique mondiale et conflits armés. Banque mondiale, FMI et GATT-OMC.* L'Harmattan, Paris: Collection La Librairie des Humanités.

Natural Resources Defence Council. (2006). Global Nuclear Stockpiles, 1945–2006. *Bulletin of the Atomic Scientists.*

OECD. (2003). *Les risques émergents au XXIe siècle.* Paris: OECD.

OECD. (2004). *L'économie de la sécurité.* Paris: OECD.

Perroux, F. (1952). *Note sur les coûts de l'homme.* Economie appliquée.

PNUD. (1999). *Rapport mondial sur le développement humain.* Paris, Bruxelles: De Boeck & Larcier s.a., p. 7–8.

Richard, T. (2009). Démocratie, économie de marché et paix. In J. Fontanel (Ed.), *Economie politique de la sécurité internationale.* Paris: Collection Librairie des Humanités, L'Harmattan.

Rutherford, M. (2009). Congressional commission focuses on China's cyberwar capability. *Military Tech.,* October 22, 2009.

Sandler, T. (2003). Collective action and transnational terrorism. *The World Economy*, *26*(6), 779–802.

Sen, A. (1981). *Poverty and famines: An essay on entitlement and deprivation*. Oxford: Oxford University Press.

Sen, A. (2002). Identité et conflit. Existe-t-il un choc de civilisations ? In J. Fontanel (Ed.), *Civilisations, globalisation, guerre. Discours d'économistes*. Collection Débats, Presses Universitaires de Grenoble.

Sheehan, M. (2005). *International security. An analytical survey* (p. 201). Boulder, CO: Lynne Rienner Publishers.

Sheehan, N. (2003). Le maintien de la paix pour le développement. In J. Fontanel (Ed.), *Civilisations, globalisation, guerre. Discours d'économistes*. Collection Débats, Presses Universitaires de Grenoble.

Sheehan, N. (2010). Gendarme mondial de la paix. Vers un nouveau rôle pour l'OTAN. In J. Fontanel (Ed.), *Economie politique de la sécurité internationale*. Collection Librairie des Humanités, L'Harmattan, Paris.

SIPRI yearbook. (2009). *Armaments, disarmament, and international security*. Oxford, Stockholm: Oxford University Press.

Smith, R. (2010). Sécurité internationale et crise économique internationale. In J. Fontanel (Ed.), *Economie politique de la sécurité internationale*. Paris: Collection Librairie des Humanités, L'Harmattan.

Smith, R., & Fontanel, J. (2008). International security, defence economics and the powers of nations. In Fontanel, J., & Chatterji, M. (Eds.), *War, peace and security* (Vol. 6, pp. 37–52). Contributions to Conflict Management, Peace Economics and Development. London: Emerald.

Stevens, B. (2004). L'économie de la sécurité émergente. Une introduction. *L'économie de la sécurité*. Paris: OECD.

Udovic, B. (2011). *Economic security: Large and small states in enlarged European Union*. Retrieved from http://www.attac.org/en/groups/center-and-periphery-abolition-neocolonialism-working-groupcpwg/bostjan-udovic-economic. Accessed on October 31, 2011.

United Nations Development Programme. (1994). *Human Development Report*. New York, NY: UNO.

Walkenhorst, P., & Dihel, N. (2004). Trade impacts of increased border security concerns. *International Trade Journal*, *20*, 1–31.

World Bank. (2003). Reducing Trading Costs in a New Era of Security. *Global Economic Prospects 2004: Realizing the Development Promise of the Doha Agenda*, pp. 179–203.

PART II
MILITARY BASES IN THE DOMESTIC ECONOMIES

DEFENCE RESTRUCTURING AND LOCAL POLICIES. THE CASE OF TARANTO, A MILITARY PORT ON THE MEDITERRANEAN SEA

Francesca Artioli

ABSTRACT

Two concurrent changes are raising questions about the interplay between armed forces and local governments in contemporary urban settings. The first one is the spatial reorganisation of armed forces that has been taking place in most European countries since the end of the Cold War. The second one is the redistribution of political authority between levels of governments that has increased the relevance of cities and transformed urban governance. The chapter conceptualises the military administration as an urban actor, whose material and symbolic resources in cities transform over time. It investigates both the effects of a (changing) military presence on urban policies, and how those changes are framed and managed by local governments. The case under consideration is the city of Taranto (Southern Italy), one among the biggest military ports on the Mediterranean Sea. Here, transformations of defence policies opened a window of opportunity for a new urban policy agenda,

The Evolving Boundaries of Defence: An Assessment of Recent Shifts in Defence Activities
Contributions to Conflict Management, Peace Economics and Development, Volume 23, 99–121

ISSN: 1572-8323/doi:10.1108/S1572-832320140000023008

whose goal is a partial differentiation from military activities. During the last ten years, local political elites have been undertaking several strategies for military spaces redevelopment. However, uneven power relations prevent civilian-military bargain: redevelopment strategies are the result of either local military initiative or central State decisions.

Do armed forces matter for urban governments? At a first sight, national defence policies and urban issues could seem far from each other. In fact, national interest and every-day urban life touch at different issues, decision-makers and organised groups. However, historians showed that urban development cannot be understood without taking into account the place of defence policies, the creation of military infrastructures and the effects of defence strategies upon the geography of urban growth. In the European urbanisation process, the organisation of military capacity and the protection of contested boundaries were driving forces for urban development and decline (Aben & Rouzier, 2001; Hohenberg & Lees, 1995).

Two parallel changes are raising new questions about the interplay between defence policies and urban governance. The first change is the spatial reorganisation of armed forces that has been taking place in most European countries since the end of the Cold War. That could be described as a double process of shrinkage and concentration (Battistelli, 1996; Irondelle, 2008). The shift to all-volunteer forces in many countries of Continental Europe, the changed nature of their missions, their Europeanisation, and their transformation into capital-intensive organisations went with the closedown of many military sites. As a consequence, armed forces have been hollowing out in many places, and concentrating in a reduced number of them, whose strategic relevance has increased. In other terms, spatial reorganisation of armed forces can be seen as a broad spatial retrenchment, where losses and gains are unevenly distributed over national territories.

The second change is the redistribution of political authority between levels of governments, along with the transformation of urban governance. Cities are now considered the core of knowledge-based economic development, and national strategies for economic competitiveness target urban agglomerations as leading places (OECD, 2006). Moreover, following different kinds of devolution policies, local governments are now responsible for increased shares of public investment, and they have to address a renewed range of policy problems, going from classical urban services to more complex programs of local development, social cohesion and responses to changing local and global economies.

The spatial reorganisation of armed forces within national boundaries and the rescaling of political authorities and economic policies in cities draw the attention on relations between armed forces and local governments in contemporary urban settings (Artioli, 2013). Scholars in economics, who calculated the impacts of defence activities on local economies, have tackled the issue. This chapter aims to bring an alternative perspective on the issue through the analytical instruments of political science and urban studies. It conceptualises the military administration as an urban actor, whose material and symbolic resources in cities transform over time. Adopting a bottom-up approach, three main questions are discussed: which are the effects of a (changing) military presence on urban policies? And how are those changes framed and managed by local governments? Which kind of urban policies are produced?

Research by urban scholars provides useful analytical tools to understand armed forces as urban actors they have widely studied relations between urban governments and those actors who control strategic resources in cities (financial and material resources, human resources and organisation, political and symbolic resources, knowledge and know-how). Most empirical research focused on local governments and economic interests, and analysed how they participate to the definition of urban policies priorities. Furthermore, scholars having addressed the evolution of city government in continental Europe showed that similar solutions have been elaborated by European local governments, in response to economic and social challenges.

One of the main changes of the last decades is the emergence of locally elaborated development projects. Here, elected authorities try to implicate those organised interests who control relevant resources in the city. Flexible forms of coordination and new partnership are established in order to deal with problems of collective action (Brenner, 2004; Harding, 1991; John, 2001; John & Cole, 1999; Pinson, 2009; Sellers, 2002).

Considering the armed forces as an actor external to urban government and having an authority upon significant resources, the chapter applies an analytical framework inspired by urban studies literature. The analysis is developed through the case study of the city of Taranto, in the touristic region of Puglia (Southern Italy). Taranto is one of the biggest military ports on the Mediterranean Sea (hosting both national and NATO structures) and headquarter of the *Dipartimento militare marittimo* 'Ionio e

Canale d'Otranto' – *MARIDIPART* (Department for the East of the Mediterranean Sea). The research was carried out in 2012. The chapter is based upon 30 semi-structured interviews with city officials (current and retired), military commanders (current and retired), and chiefs of civilian and military administrations. Moreover, the main local newspaper, *La Gazzetta del Mezzogiorno*, and official city documents concerning urban public policies have been analysed.

Two main arguments are developed. The first is that transformations of defence policies open a window of opportunity for a new urban policy agenda, whose goal is a partial differentiation from military activities. During the last ten years local political elites have been undertaking several strategies for the transformation of military spaces. The second argument is that uneven power relations and the political weakness of Taranto's local government prevent the institutionalisation of local civilian-military bargain for urban reconversion. As a consequence, redevelopment strategies are a result of local military initiative, on the one side, and of State decisions, on the other side.

UNDERSTANDING MILITARY PRESENCE THROUGH THE URBAN LITERATURE

Theoretical Insights

The main literature this chapter refers to considers relations between groups, economic interests and local policies. Three main trends of research help to conceptualise armed forces as a changing urban actor.

(1) The European sociological tradition in urban studies, stemming from a question about how social and political orders are constructed, analyses city governments and social regulation processes in cities specialised in some activities (one-company towns, city-ports, etc.). The starting point of this research is a general analysis of the influence of particular activities to determine the social structure of the city (Bagnasco, 1986; David, Duriez, Lefebvre, Voix, & Tellier, 2006; Dormois, 2008; Pizzorno, 1960). Then, they focus on regulation, defined as *coordination* of different activities, *resource allocation* and *conflict prevention and solving* (Lange & Regini, 1989).

Rescaling social analysis from national State to cities, they empirically demonstrate that different kinds of activities in cities entail different arrangements in urban policies and politics. In other words, urban policies and social and political regulation of a city whose development is closely linked to a specific activity (a port, a large company) cannot be understood without taking into account, through a long-term analysis, groups related to this activity, their elites and the influence they exercise. In some terms, these authors therefore claim for the uniqueness of social and political arrangements and for the need to differentiate among different urban actors.

(2) The political economy approaches to urban policies elaborated theories in order to explain interactions between the urban political sphere and economic interests. These approaches help focus on those actors who control relevant resources in cities and conceptualise their influence on local policy-making.

A first group of scholars interprets urban policies as the result of class struggles and capitalist laws (Harvey, 1985; Logan & Molotch, 1987). In Marxists approaches to local government, *policy* plays a very little role and a stronger attention to the political sphere and to the wider city *polity* has been attributed to the tenants of the *regime theory*. Although they still focus on capitalist society, they take into account a wider range of actors than economic interests, and they consider issues related to political competition. In fact, *regime theory* focuses on relations between governmental and non-governmental actors in cities in order to determine stable ruling coalition. A regime is therefore an informal yet relatively stable group with access to institutional resources that enable it to have a sustained role in making governing decisions (Stone, 1989). The empirical research focuses on processes of coalition building, interactions among actors and bargaining contexts.

Since these mentioned theories share the United States as a common origin, their applicability in the European context has been strongly debated (Harding, 1997; Stoker & Mossberger, 1994). Indeed, considering European cities, some elements have to be taken into account in order to prevent a non-reflexive theory from being taken out of context. These elements are the weaker place of firms in local politics (mostly in endorsing elections candidates), the smaller relevance of tax incomes in local budgets because of public transfers from the central government, and above all, the higher role of the central State in determining local policies, the weight of de-concentrated administrations of the central State within the cities, and the importance of public landownership which reduces the amount of urban land available in the market (Kazepov, 2005).

(3) Finally, scholars having addressed the evolution of European city governments since the 1970s help formulate hypothesis both about changes in *how* European cities are governed and recurrent *policy outcomes*. Considering the first point, scholars pointed out a process of *redistribution of political authority* between levels of governments and blurring boundaries between *private and public actors*.

These changes are accounted to devolution followed by European centralised States and to the evolution of the role of the State in the globalised capitalism (Brenner, 2004; Brenner & Theodore, 2002; Le Galès, 2003; Pinson, 2009). Many European countries have experienced reforms in their institutions of local governments by adding power of sub-national elected authorities and transferring responsibilities to sub-national political units (Bobbio, 2002). Further, a change in State policies addressing cities' problems and local development has undermined the redistributive function of territorial policies in Western Europe. Until the 1970s, the State was the main actor ruling economic and planning policies in cities. But, the shrinkage of State resources and the limitations to economic interventionism, introduced notably by the European Union, entailed the displacement of urban policy production to cities themselves. Territorial policies have more and more focused on promoting local actors' mobilisation rather than focused on redistributive issues or direct support to economic development. In a context of limited public resources, cities have to elaborate their own projects in order to compete for the allocation of central State resources (Béhar & Estèbe, 1999; Pinson, 2009).

Furthermore, economic change has intensified pressures on municipalities. Large enterprises and public-sector organisations that used to provide stable employment in many localities have been restructured. The shrinking of traditional industries and the increased mobility of capital and labour put cities in competition to attract 'factors of production' (Scott, 2008). In this context, the needs of *locally-elaborated development projects* to get financial transfers from the centre went with the increase of number and types of actors implied in urban policies. Here, city officials have to mobilise actors controlling relevant resources (firms, universities, local elites) and to build up general and shared policy agendas, along with new institutional venues for policy bargain (Pinson, 2009). Their role in aggregating interests and preventing gridlocks in policy implementation has increased, which could reveal a manner to increase their political authority and enhance legitimacy and recognition (Borraz & John, 2004).

> These changes in social and political regulation have been conceptualised as a shift from urban government to urban governance. The latter can be defined as a process of coordinating actors, social groups and institutions to attain particular goals, discussed and defined collectively in fragmented and uncertain environments (Le Galès, 1995). Thus, key elements are the densification of information exchange networks, the recognition of shared interests between actors and the consolidation of a common framing of policy problems (DiGaetano & Strom, 2003; Kilburn, 2004; Pierre, 2005; Stoker & Mossberger, 1994).

Considering policy outcomes, similar policies of urban redevelopment have been implemented in response to the changes of urban and national productive systems. Land uses in European cities have massively changed during the last decades. Those spaces which resulted from massive investment in industrial and transportation systems and have become brownfield were reinvested by urban policies in similar manners (Marcuse & Van Kempen, 2000). A prominent example of this phenomenon is the process of waterfront renewal in industrial and port-cities. As a consequence of changing technology and economy of global shipping, shipping activities have been either delocalised outside city centres or closed down.

Massive public and private investments have been dedicated to the transformation of port and industrial brownfields, aiming to increase land values and develop touristic and leisure activities, select housing and office buildings (Marcuse & Van Kempen, 2000). Born in Western industrial cities, policies of waterfront renewal have diffused as successful urban development agenda, fostered by widely known cases such as Bilbao. As Hoyle argues, 'almost every city with any form of water frontage-at least in advanced countries-is doing something about revitalizing its waterfront, if such renovation can be considered remotely affordable and if the essential political impetus is present. This process involves not only port cities, but all kinds of other cities: on lakes, rivers, canals, and artificial water bodies' (Hoyle, 2000, p. 387).

To put it shortly, in the remaking of the global political-economic space, cities have played and have been assigned a pivotal role for economic growth. Despite tensions and struggles, cities have come to be considered as collective actors having institutional and economic capacities both for innovating and for elaborating responses to State retrenchment and to the readjustment of their productive systems (Le Galès, 2003). The 1990s and the beginning of the 2000s can therefore be seen as a 'golden age' for urban resurgence and urban governance. Empirical research focused on the processes allowing actors' resource mobilisation, paths leading to the emergence of bargaining venues for collective action in cities, and policy outcomes in urban renewal.

Formalizing Armed Forces as an Urban Actor

Drawing on the discussed literature, it is possible to define some features of the military presence in the city and formulate guiding hypothesis about change in a military city port. Analysing the armed forces in the city requires considering them under three different and complementary aspects.

Armed forces are a State institution and organisation. They can be defined as a complex institutional organisation in charge of managing the monopole of the legitimate violence (Hamelin, 2000). Even if the internal coherence of military organisation should not be over-estimated, they have specific goals, a range of priorities and interests to defend and implement, and patterns of decision-making. Approaching them from the perspective of local governments means that decision centres are located at different institutional levels. Elected city authorities have their legitimacy and accountability at city level, and they benefit of some autonomy for policy decision and implementation. On the contrary, armed forces at the local level are a public administration, which is part of a hierarchical system whose heads are located in the capital or at least at the regional level.

Armed forces in the city can be understood as a resource system. The organisation controls: material resources (real estate, public funding), political resources and social legitimacy (because of its function of defending the nation), and knowledge and know-how (because it is specialised in a specific range of activities). The resource system is neither coherent nor unchangeable: organisation's resources transform over time and they are linked to political and budgetary decisions, technological transformations and the evolution of informal norms and ways of thinking.

Shifting from the organisation to individuals, armed forces are a group of workers, urban dwellers and voters. In this realm, the end of conscription has modified the situation, since needs, living-standard expectations and the familiar constraints of volunteer and professional servicemen are higher than those of drafted soldiers. Since the 1990s, the Italian MoD has to deal with new issues related to the management of human resources in military bases (Fig. 1).

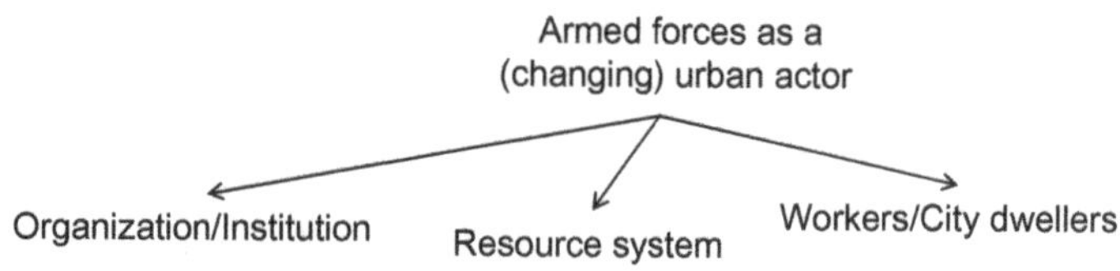

Fig. 1. Understanding Armed Forces in Urban Governance.

Armed forces are here considered as a collective actor external to urban government and having an authority upon significant resources. On the one side, their activities and demands can influence ways local decisions are taken and the content of urban policies. On the other side, their presence can be more or less addressed in policy framing and governed by elected city authorities, following transformation in available resources, evolutions of power relations or in systems of incentives and constraints for policy-making.

For this reason, the urban effects of a military presence cannot be considered only as the effects of an exogenous variable over a system 'taken for granted'. Reality is far more complex: local regulations and political activities of municipalities are critical for the analysis of urban outcomes. The empirical research therefore has to focus on power relations, resource exchanges and their transformations over time.

This conceptualisation of armed forces as an urban actor and the discussed literature allow formulating the main hypothesis that guides empirical research. It can be argued that cities with a longstanding military presence are characterised by specific patterns of resource allocation and conflict solving, determined by long-term interactions between military activities and urban policies. Furthermore, it is likely to observe changes towards the elaboration of reconversion strategies similar to those accounted for other European cities.

TARANTO: CHANGING MILITARY ACTIVITIES IN A SPECIALISED CITY

Taranto as a Military and Industrial City

How the Navy and Ministry of Defence control resources in Taranto must be clarified in order to understand power relations and the ways issues are perceived and framed in policy-making. The decision to locate in Taranto one of the arsenals of the newly born Italian State was taken in 1882 thanks to the mobilisation of local elites, who succeeded in securing public investments in the city against alternative locations for the new infrastructure. This choice has determined successive steps in Taranto urban growth, opened a phase of intense urban development and radically modified urban layout (Petrilli, 1970).

The military base and the arsenal were inaugurated in 1887. In the context of the rural Southern part of Italy, they were symbols of modernity and progress. The establishment of a military arsenal and port was followed by a first demographic boom. Between 1881 and 1921, the population grew at rates comprised between +15% and +45% a year (Camera di commercio industria artigianato e agricoltura di Taranto, 1977). The city became an attractor for other Italian Southern regions, and its relative demographic weight grew sharply: if in 1891 Taranto population was 1.82% of the population of Southern Italy, by 1951 it has raised to 3.67% (Petrilli, 1970). Considering the social composition of the city, this national choice also entailed the emergence of an urban and industrial working class. The Second World War is the moment of greatest activity of the arsenal. Between 1936 and 1951, urban population grew by 51,000 inhabitants, and 13,000 civilians were working for the Ministry of Defence (the highest number in the history of the military arsenal).

This period was followed by a first decreased of military activities, and by the emergence of the second specialisation of the city, which has become in time the prevailing sector in Taranto. In 1959, the central government decided to locate in Taranto one of the four nationally owned steel-producing centres Italsider (*IV Centro Siderurgico Italsider*). Again, the intervention of the central State brought to Taranto massive public investment. The location in Taranto was both due to political reasons; local political elites put forward their ability in lobbying the national government. The project is to be understood as part of national policies for the industrialisation of the South, but also 'in a global perspective for the national interest [...] whose concerns were far beyond territorial policies for the industrial growth' (Petrilli, 1970, p. 1551).

Like the period following the creation of the military arsenal, this second phase of city development was considered as a modernising opportunity in an underdeveloped South. The planning documents for the steel industry were realised by well-known progressive engineers, and claimed as an 'example of territorial planning' for the country (Tekne, 1965). The demographic growth of the 1970s is explained by this activity: between 1961 and 1981, Taranto population grew of 49,000 people.

The steel crises beat the city hardly, and, after the privatisation of Italsider in 1991, the military arsenal played, for a short time, the role of absorbing unemployment caused by steel restructuring.

The city expanded almost uninterruptedly for 90 years as a consequence of the two State-led industrial activities and was followed by a steep decrease of urban population in recent times, due to industrial crises and the harsh pollution and public health problems of the metropolitan area. The particular model of Taranto development is therefore closely linked to national strategies and to State-led activities. These latter ensured a redistributive function thanks to military and industrial public programs (with negative effects during phases of expenditure cuts and of steel crises) and had both strong spatial effects and environmental consequences.

Taranto economic and political elites continuously put forward the problem of the structural disequilibrium of the local economy, dependent on public expenditure and with poor endogenous firms and entrepreneurship. Innovation and modernisation in Taranto were always endogenous and high-skilled workers to be found mostly only in State-led activities. Additionally, Taranto urban layout is structured by massive fixed investments in military and productive systems, which are radial points for surrounding developments and an uncontrolled process of urbanisation. The arsenal and the military base are both located nearby the historical city centre (a small island in the bay), and they originate the first phase of urban expansion on the East side of the bay. Indeed, a new neighbourhood (which is today part of the city centre) developed around the walls of the arsenal and military base, and occupied entirely the Southern side of the smallest bay (site no. 5 in Fig. 2). The arsenal surface is around 90 hectares, half of the urban waterfront of the city centre is inaccessible because of military activities and hidden to the view by a continuous wall of 3,250 kilometres length and 6 metres height (also called in Taranto *The Great Wall* (Camera dei Deputati – Servizio Studi, 1992)). The need for a reorganisation of military spaces has been a constant point in local political debate, either inspired by the needs for a better matching between arsenal infrastructures and activities, or by the hopes for alternative activities to be developed in the city.

Importantly, the economic and spatial weight of national (or formerly national) sectors contributed to structure the position of Taranto elites *vis-à-vis* the central State. On the one side, they advocate the capacity to attract massive national projects and investments, whose locational choices were in competition with other southern cities. On the other side, they sustain a representation of the city of being dependent on decision taken in Rome, which makes the city subdued to national industrial and military strategies that do not really take into account the well-being of Taranto inhabitants and the local consequences of industrial and military restructuring. This double register structures the relations between urban political elites and the State and it is the cognitive framework any agenda for urban development is embedded in (Fig. 2).

Dynamics of Change: Functional Concentration and Spatial Decentralisation of the Navy

This section examines the dynamics of change of military activities (the Navy and military arsenal) in the city. The biggest military infrastructure realised in Italy after the end of the Second World War has been

Fig. 2. Urban Space under Military Control in Taranto.

constructed in Taranto. It is the new military base (site no. 8 in Fig. 2), located on the waterfront of the biggest and deeper bay. The opportunity of moving the military base from the smaller to the bigger bay was part of a project elaborated under the fascist regime. The idea came back to public debate in the 1970s and was finally approved in 1985 and completed in 2004. NATO financed around 50% of this investment. Italian Navy warships and other NATO forces have moved to the new location. This infrastructure was considered as necessary because of the increased size of warships and of the increased strategic relevance of Eastern Mediterranean Sea, at the expense of other locations of the Italian Navy (in particular ports of La Spezia and Naples). In spatial organisation, Navy core activities have been decentralised outside the city centre, in a neighbourhood of the urban fringe. However, the construction of the new base did not automatically entail the disposal of those parcels behind the *Great Wall* that remain under military authority.

The second dimension of change is the decline of arsenal activities and workforces. The lack of qualified workforces and the ending of national programs of workforce training, along with the diminished number of warships in Italian Navy and the increased role of defence industries in military procurement (*Finmeccanica, Fincantieri*), led progressively to an excess of spaces and workforce in the arsenal. Called 'the problem of empty docks', this entailed the need for public investments for re-adaptation of existing structures. Therefore, there was a continuous decrease in the number of civilian workers in the military arsenal. They were 4,774 in 1975 (Camera di commercio industria artigianato e agricoltura di Taranto, 1975), 2,935 in 1993 and 1,503 in 2011 (data from the Ufficio Contabilitaè Statistica Arsenale Militare di Taranto). This decline has been accompanied by several mobilisations from labour unionists and local politicians, claiming for new training programs and subsequent hiring.

Table 1. Ministry of Defence Employees in Taranto (2011).

Military employees	9,648[a]	– 1,203 officers – 5,805 non-commissioned officers – 2,820 non-commissioned members[b]
Civilian employees (2011)	3,771	
Total	13,419	
Taranto inhabitants	191,810	

Sources: MARIDIPART and ISTAT.
[a]About half of them are on board.
[b]Engaged for 1 year (VFP1) or 4 years (VFP4).

In contrast with naval maintenance activities, military personnel and activities followed a different pattern of change (Table 1). Indeed, both the construction of the new military base and the changing geopolitical landscape have strengthened the military function of the city. In 1985, seamen were around 9,000 along with 1,000 draftees. This has not radically changed because 9,648 professional seamen are employed today in Taranto and the Ministry of Defence injects in the local economy around 310 million euros per year (wages and procurement).

The trajectory of change of the Navy in Taranto can be seen as a movement of decentralisation (from the city centre towards the peripheral neighbourhoods) and of functional concentration, with the reduction of warship maintenance in favour of the role of national and NATO military base on Mediterranean Sea. The suspension of the draft (decided in 2001) and the transfer of the military base were almost concomitant. As a consequence, several spaces have remained completely empty, while other military offices are scattered in quasi-empty sites.

Following these changes, the Navy engaged an internal assessment about its spatial organisation in the city. The department in charge of the infrastructures elaborated the so-called 'Navy urban plan', which was released in 2011. The underlying logic was both financial and functional. On the functional side, there was the need for finding a better functional correspondence between occupied spaces and current military functions and needs through the disposal of the majority of assets in the centre (apart from the more symbolic ones such as the siege of the Military commander for Eastern Mediterranean Sea) and the concentration in a reduced number of sites outside the centre. On the financial side, resources were expected from sales of military lands to private or public actors, which should contribute to finance both the transfer outside the city centre and the construction of housing for professionalised seamen.

AN IMPOSSIBLE URBAN RENEWAL? PROBLEMS IN URBAN COLLECTIVE ACTION

This section considers the kind of political relations and bargaining processes at the urban level. It shows difficulties encountered by Taranto elected officials in dealing with the transformation of defence policy sector. Changes in military activities can be considered as comparable with those accounted for big civilian ports, where shipping activities have been either delocalised outside the city centre or closed down. Following the urban literature, Navy transformations could have generated projects of waterfront renewal similar to those accounted for comparable port and industrial cities.

In Taranto, waterfront renewal through the transformation of the old military base is at the heart of any project for urban reconversion and economic differentiation. On the one hand, changing land uses would allow for the development of new activities benefiting from the central location of the land, in the heart of the city, where institutional centres and commercial facilities are currently located. On the other hand, urban projects have the double function of effectively changing the use of land assets while symbolising the urban renaissance and the effectiveness of political elites. Indeed, land use and investment are as much related to the symbolic dimension of power as to economy (Zukin, 1991).

Therefore, land controlled by the Navy is a key resource for the implementation of any kind of agenda aiming to develop cultural and touristic activities in the city. However, the fragmentation of the governance system, the absence of a coalition for local development and the poor financial capacity of the city government make urban renewal a difficult task. In particular, projects of urban reconversions based on waterfront renewal have never been implemented.

The Impossible Redevelopment of Military Waterfront

The spatial decentralisation of military activities generated high expectations from Taranto elites and civil society. Since the decision of constructing the military base, the issue of waterfront renewal has become a recurrent theme in local political debates. The issue was first addressed at the beginning of the 1990s by a deputy-mayor, who was also labour union leader in the military arsenal. Mobilising his networks within the Ministry of Defence, he elaborated a first attempt to conclude an agreement with the Ministry for the disposal of the old military base on waterfront. However, in the context of the Italian political instability of the 1990s, the local majority changed several times and the project was put aside.

It re-emerged in 1999 and in a more articulated manner, when a local politician was appointed Secretary of Defence in a centre-left coalition government, and continued when, in 2005, he was charged of tourism in the Regional government. For the first time since decades, a Taranto politician was holding a governmental post, which gave more direct access to the military administration. Indeed, he had political resources not only for dealing with the Navy at the local level in Taranto, but also for sustaining local demands to the central State. The middle of the 2000s was characterised by the elaboration of a project for redevelopment of military parcels, consisting in the reconversion of the main buildings on the shore, and eventually in the transformation of the dismissed aircraft carrier Vittorio Veneto, currently abandoned in the old military port, into a museum of naval activities. Several local actors from the organised civil society federated around this project, including the provincial Architects Association, a private foundation for naval activities (*Fondazione Ammiraglio Michelagnoli*) and other professionals.

The project was based on the idea of waterfront renewal for touristic purposes and was broadly inspired by well-known projects of waterfront renewal, both in Italy (Genoa) and abroad (Bilbao). During this period, and under the impulsion of the Regional Department for Tourism led by the former Secretary of Defence, they generated knowledge and ideas for reconversion and built up new linkages which stabilised over time. However, the centre-right majority at the local government never supported the project, leaving other social and political actors very few help in the process. In 2006, the project was put aside another time because of the financial distress of the city, which contributed to marginalise any kind of development policy.

Local Collective Action: Uneven Power Relation and Weak Bargain Venues

Despite the decreasing role of the arsenal and the transfer of the main military base to a city peripheral neighbourhood, no coherent policy for the reconversion of the sites and for the renewal of urban identity has been implemented in Taranto. The weakness of local government has prevented any proper negotiation for the dismissal of the old military base in the city centre and for surrounding buildings (for instance, the great Officer Circle).

There is no local coalition for urban development, and no routinized and institutionalised venue for civilian-military negotiations. In this respect, Taranto differentiates from other military and industrial cities, where city officials elaborated strategies to mobilise those actors controlling relevant resources, to build up general and shared policy agendas, and to create local institutionalised decision venues for bargain (Artioli, 2013). Given power relations between State military institutions and local governments, local institutionalised venues for bargain serve to reduce uncertainties and to create a sense of equality between both parties (Artioli, 2013). The absence of urban coalition for local development in Taranto is due to the weakness of the local government in aggregating powerful actors around a common development agenda.

First, in Taranto there is no shared understanding of urban problems and policy priorities because of the existence of two predominant activities (steel industry and the Navy) putting local government under dependence from the two main institutional and economic actors involved in two different sectors. Secondly, there are no stabilised relations. A relevant example of this situation is the process for elaborating a strategic planning document that identifies the most relevant actions to be undertaken for the metropolitan area and aiming at aggregating actors around a common project. The project was launched by the regional government in order to select a number of projects which should benefit from regional targeted grants. The first phase of the process required several thematic workshops providing meeting venues for all relevant actors. However, both the Navy and ILVA (the steel firm that resulted from Italsider) were met separately from other local actors and aside from the general consultation process. As argued by the civil servant responsible for the strategic planning, 'they do not seat around the table with others (local actors). Separate meetings are organised. The Navy and ILVA are always there, but you do not see them in these kind of meetings'.

Perceived and real uneven power relations explain local difficulties in dealing with these actors. As a consequence, all local civilian officials share the idea that a diversification agenda is to be found through negotiations with the central government, considered as the only actor having sufficient resources for dealing both with the military administration and the steel company. Indeed, the absence of local bargain where the Navy could be considered as a partner in local government actions is accompanied by the idea that any kind of agreement is to be concluded with the State and thanks to political relations.

Three key interventions are expected from central and regional governments: investments for the depollution of bay waters and urban soils, deeply impacted by one hundred years of highly-polluting and softly-regulated industrial activities; investments for the reconversion of military lands; and finally the development of the industrial port, currently dependent on the upside-downs of steel industry. Importantly, this agenda does not consider either ILVA or the Navy as possible partners, but tries to design an alternative urban future despite, and beyond, their presence.

BEYOND THE NAVY, WITH THE NAVY

Military Initiatives for Urban Development

As a result of weak capacities of the local government and uneven power relations, projects between the city and the Navy are currently realised under the impulsion of specific personalities.

The most important initiative is the reconversion for touristic purposes of the highly symbolic building *Castello Aragonse*, currently the most-visited touristic attraction of this province. The fortress is located in the core of the city centre, beside the old bridge linking the historical city (built before the 19th century) and neighbourhoods whose development started after the unification of Italy (1861) and the decision to locate in Taranto a military arsenal of the newly-born State. The fortress dates back to the 15th century and since the end of the Second World War was used as a healthcare centre for draftee seamen. The historical structure was completely hidden by modern infrastructures serving the organisation of draft. After the suspension of the military service, the place lost its function and was abandoned, despite its historical and symbolic value.

The reconversion of the site was the initiative of a former MARIDIPART commander, an admiral who retired in 2005, passionate about archaeology. The personality and the specificity of interests of this former commander made possible to renew city-navy relations for a specific realm. Indeed, the building is still a property of the Ministry of Defence and mobilises skills and resources from the military arsenal (metallurgy, engineering, etc.) to achieve fortress restoration. Despite the exceptional and contingent configuration which gave origin to this touristic initiative, the fortress restoration was put forward by civilian authorities as the first step of a new history for the city, still to be written. On the other side, the Navy uses it as a form of legitimisation of its presence in the city and to demonstrate military involvement and concerns in Taranto development.

A second project implemented with the Navy is related to a shared use of the sports hall located within the arsenal and which was formerly used by draftees and civilian arsenal workers. The structure was underemployed by the military administration and thus it was falling into disrepair. The initiative to reach an agreement in 2008 between the provincial government and the Navy was undertaken by an elected official of the Provincial Council, also an employee in the military arsenal (and labour unionist). Importantly, he was helped by a retired admiral, who performed the role of being his advisor for the relations with the Navy in Taranto.

Despite differences between them, these projects constitute an attempt to go beyond the industrial and military specialisation of Taranto and to mobilise land and buildings owned by the military administration for alternative purposes. Importantly, members of the military administration played a supportive role in all these processes. Indeed, labour unionists of arsenal and retired admirals were *gatekeepers* between local political and administrative elites and the military administration. Despite differences between them, they are 'reducers of uncertainty' (Crozier & Friedberg, 1980), because they belong two both worlds and they represent part of the military administration when interacting with local politics and part of local politics when interacting with the military administration.

State Responses to the Urban Crises

Unsuccessful negotiations for waterfront disposal and punctual projects for the fortress and the sports hall are two alternative models of civilian-military negotiations in urban policies, mostly dealt with at the local level, but in a highly fragmented system and in the absence of stabilised venues for bargain.

A third kind of policy negotiation is related to new claims put forward by municipal and provincial governments towards the State, framed as a compensation for sacrifices to the national military and industrial interest which have structured the city development and its current problems. Indeed, the city faces a dramatic health crises related to the abnormal level of dioxin in air and soil, whose damages on inhabitants' health and on the environment have been proved by a judicial inquiry. These are due to steel activities, one of State activities which structured Taranto development over time. Despite the fact that the steel company has been privatised in the 1990s, the State is the former owner and therefore considered as co-responsible for the current crises. New relations between the city and the Navy have therefore to be understood in this context of renewed relations between the State and local government, which opened the way to the establishment of new stable relations with local government (a negotiation table for the future of the metropolitan area).

The historical dependence of the city from exogenous decisions has become an argument for new claims towards the central State. Indeed, new demands elaborated by local public actors and the organised society are called a new 'season of compensations for the recovery and development of the area' (Provincia di Taranto et al., 2012). In this respect, Taranto has to become a national priority for depollution and development because 'during the last 150 years (*from the constitution of the Italian State*) no other Italian territory has been the object of such massive and invasive activities, such as the military arsenal, shipbuilding, steel industry and petrochemical industry, and which have caused heavy effects on the territory and the environment' (Provincia di Taranto et al., 2012).

This claim is translated into several demands for State intervention, going from depollution to infrastructural investments for port development and urban regeneration. The disposal of Navy lands is part of projects for urban renewal, and it has been finally accepted in January 2013 by the central government (Ministry of Defence, Ministry of Finance). This can be considered as a turning point for the relation between Taranto and the Navy, because it opens the way to further spatial reorganisations in the city. Additionally, parcels will be disposed for free, as an exception due to incidental crises and in contradiction with what was expected by the Navy in terms of financial resources. State-city negotiations overwhelmed military expectations.

However, lands concerned by the disposal are around the military arsenal, but do not include the waterfront or other buildings within the wall. Indeed, they have been chosen because of their easy availability and the existence of local projects for re-use. Empty parcels are to be converted in parking lots, former military housing will be used to face local housing crisis and the former leisure centre for arsenal employees into a new auditorium for arts and theatre. The resulting policy process, therefore, does not follow the classical patterns of urban reconversions and regeneration, and it is more concerned with urgent local needs.

CONCLUSION

The chapter aimed to analyse the effects of military presence on cities. Instead of focusing on economic effects of defence expenditure on local economy, it aimed to bring some elements of understanding of the effects on policy-making and to position them in the context of contemporary urban change.

The chapter proposed an interpretative framework inspired by theories of urban studies, and considered the military presence as an institution, a system of resources and a specific population. Adopting a bottom-up approach, the in-depth study of Taranto focused on power relations and resource exchanges between the Navy and local government. It shows that changes in military activities can be considered as comparable with those accounted for big civilian ports, where shipping activities have been either delocalised outside the city centre or closed down. Local political elites have been elaborating several projects for the transformation of military spaces. However, the extreme fragmentation of local actors and uneven power relations prevent the institutionalisation of local civilian-military bargain for urban reconversion.

As a consequence, redevelopment strategies are the result of local military initiatives from retired admirals and labour unionists in the military arsenal, and of punctual agreements with the central State in reaction to the heavy cries of the city. In this respect, Taranto differs from other European military cities where local government has push forward the need for establishing stabilised negotiations with military authorities in a broader agenda of differentiation from military activities.

REFERENCES

Aben, J., & Rouzier, J. (2001). *Défense et aménagement du territoire: Colloque de Montpellier, 4 et 5 décembre 1997*. Montpellier: Publications de l'Université Paul-Valéry.

Artioli, F. (2013). The navy and the city: Conflict, cooperation and political competition in the urban governance of Toulon. *Urban Research & Practice*, *6*(1), 75–94.

Bagnasco, A. (1986). *Torino. Un profilo sociologico*. Torino: Einaudi.

Battistelli, F. (1996). Soldati: Sociologia dei militari italiani nell'era del peace-keeping (Collana di sociologia militare), F. Angeli.

Béhar, D., & Estèbe, P. (1999). L'Etat peut-il avoir un projet pour le territoire? *Les Annales de la Recherche Urbaine*, (82), 80–91.

Bobbio, L. (2002). *I governi locali nelle democrazie contemporanee*. Roma-Bari: Laterza.

Borraz, O., & John, P. (2004). The transformation of urban political leadership in Western Europe. *International Journal of Urban and Regional Research*, *28*(1), 107–120.

Brenner, N. (2004). *New state spaces: Urban governance and the rescaling of statehood*. Oxford: Oxford University Press.

Brenner, N., & Theodore, N. (2002). Cities and the geographies of 'actually existing neoliberalism'. *Antipode*, *34*(3), 349–379.

Camera dei Deputati – Servizio Studi. (1992). Relazione n. 813 sulla visita di due delegazioni della Commissione Difesa della Camera dei Deputati all'arsenale di La Maddalena, all'Arsenale di Taranto, alla Stazione Navale e alla Scuola Truppe Corazzate di Lecce, p. 21.

Camera di commercio industria artigianato e agricoltura di Taranto. (1975). Atti della tavola rotonda sul tema 'Lo sviluppo di Taranto tra l'Arsenale e il IV Centro Siderurgico' del 15 gennaio 1975 Supplemento a Produttività Ionica, vol. 6 anno V. Taranto: Camera di commercio industria artigianato e agricoltura di Taranto.

Camera di commercio industria artigianato e agricoltura di Taranto. (1977). Taranto Progetto 80, Progetto di Sviluppo Metropolitano della Provincia Ionica C. Pace, éd., Franco Angeli.

Crozier, M., & Friedberg, E. (1980). Actors and Systems: The Politics of Collective Action. University of Chicago Press (Tx).

David, M., Duriez, B., Lefebvre, R., Voix, G. & Tellier, et T. (2006). *Roubaix: Cinquante ans de transformation urbaine et de mutation sociale.* Villeneuve-d'Ascq: Presses Universitaires du Septentrion.

DiGaetano, A., & Strom, E. (2003). Comparative urban governance: An integrated approach. *Urban Affairs Review*, *38*(3), 356–395.

Dormois, R. (2008). Les coalitions dans l'analyse des politiques urbaines post-keynésiennes. Discussion à partir de la comparaison des politiques de régénération urbaine menées dans trois villes européennes en reconversion. *Métropoles*, (4). Available at http://metropoles.revues.org/3122

Hamelin, F. (2000). *La spécificité militaire, une construction institutionnelle: Le cas de la formation des officiers du Génie au XIX et XX siècle.* Paris: Institut d'Etudes Politiques.

Harding, A. (1997). Urban regimes in a Europe of the cities? *European Urban and Regional Studies*, *4*(4), 291–314.

Harvey, D. (1985). *The urbanization of capital: Studies in the history and theory of capitalist urbanization.* Baltimore: The Johns Hopkins University Press.

Hoyle, B. (2000). Global and local change on the port-city waterfront. *Geographical Review*, *90*(3), 395–417.

Irondelle, B. (2008). *Les politiques de défense. In Politiques publiques 1, La France dans la gouvernance européenne* (pp. 93–112). Paris: Presses de Sciencies Po.

Kazepov, Y. (2005). *Cities of Europe: Changing contexts, local arrangement and the challenge to urban cohesion.* Oxford: Wiley-Blackwell.

Kilburn, H. W. (2004). Explaining U.S. urban regimes: A qualitative comparative analysis. *Urban Affairs Review*, *39*(5), 633–651.

Lange, P., & Regini, M. (1989). *State, market and social regulation: New perspectives on Italy.* Cambridge: Cambridge University Press.

Le Galès, P. (1995). Du gouvernement des villes à la gouvernance urbaine. *Revue française de science politique*, *45*(1), 57–95.

Le Galès, P. (2003). *Le retour des villes européennes.* Paris: Les Presses de Sciences Po.

Logan, J. R., & Molotch, H. L. (1987). *Urban fortunes: The political economy of place.* Berkeley: University of California Press.

Marcuse, P., & Van Kempen, R. (2000). *Globalizing cities: A new spatial order?* Oxford: Wiley-Blackwell.

Petrilli, G. (1970). *Lineamenti dell'evoluzione socio-economica della città di Taranto. In Scritti in onore di Gaspare Ambrosini* (pp. 1533–1554). Roma: Pubblicazioni della Facoltà di Giurisprudenza dell'Università di Roma.

Pierre, J. (2005). Comparative urban governance: Uncovering complex causalities. *Urban Affairs Review, 40*(4), 446–462.

Pinson, G. (2009). *Gouverner la ville par projet: Urbanisme et gouvernance des villes européennes*. Paris: Les Presses de Sciences Po.

Pizzorno, A. (1960). *Comunità e razionalizzazione*. Torino: Einaudi.

Scott, A. J. (2008). Resurgent metropolis: Economy, society and urbanization in an interconnected world. *International Journal of Urban and Regional Research, 32*(3), 548–564.

Stoker, G., & Mossberger, K. (1994). Urban regime theory in comparative perspective. *Environment and Planning, 12*(2), 195–212.

Stone, C. N. (1989). *Regime politics: Governing Atlanta, 1946–1988*. Lawrence: University Press of Kansas.

Tekne. (1965). Il piano di Taranto. Il piano regolatore territoriale dell'Area di Sviluppo Industriale di Taranto.

Zukin, S. (1991). *Landscapes of power: From Detroit to Disney World*. Berkeley: University of California Press.

THE LOCAL FOOTPRINT OF DEFENCE INDUSTRY: THE NATURAL EXPERIENCE OF BREST

Thierry Sauvin and Roland de Penanros

ABSTRACT

In situation of uncertainty, many companies refuse today to bind their destiny to the fate of a specific territory. Nevertheless territories are engaged in trajectories that result from their industrial history and legacies. Companies are significant actors in structuring a given territory. Their territorial footprint is plural: industrial, organisational, psycho-cultural, environmental and political. It creates trajectories, hysteresis effects and irreversibility. We will mobilise cognitive approaches and more singularly the evolutionary approach to appreciate the territorial print of a given company. We focus on the Defence industry in Brest region (Brittany, France), a singular territory formatted by the naval shipyard industry. This latter has been involved in deep reorganisations since the beginning of the 1990s. This leads to many questions about possible industrial diversification and how to rebuild a cognitive trajectory of such territory by involving local authorities and the civil society.

The Evolving Boundaries of Defence: An Assessment of Recent Shifts in Defence Activities
Contributions to Conflict Management, Peace Economics and Development, Volume 23, 123–138

ISSN: 1572-8323/doi:10.1108/S1572-832320140000023009

Facing strong pressures of financial markets, companies have initiated, since the beginning of the 1980s, permanent reorganisations (Raveyre, 2005). Territories, of which some sheltered these companies during several decades, are today handicapped by the closing-down of establishments and subsidiary companies, outsourcing and offshoring. Although rich in competences, these territories, at least for a number of them, remain locked in yesterday's competencies and have difficulties to move forward to other trajectories of development.

The idea raised by this chapter is that territories are deeply marked by industries that they welcome. The company is not an insignificant actor but a structuring element in the local economic systems. This is particularly the case when analysing Brest (France), a city that has been developing around a naval shipyard and French Navy activities. Indeed, Brest is a natural harbour chosen in the 17th century for housing part of the fleet as well as an arsenal, which was privatised a decade ago (Direction des Constructions Navales System [DCNS]). Today, Brest is the Navy base for deterrence submarines. This area represents a natural experience of the relationship between a large company that is a leader in naval systems and the territory in which it is located.

The territorial footprint of the company is plural. It is industrial, organisational, psycho-cultural, environmental and political. The production of physical and human resources and specific assets leads to irreversibility, inertias preventing or slowing changes and diversification. The second concern is that the more one waits for adjusting, the more competences held by actors evaporate. The capacities of actors to be and act, within the meaning of Sen (2000), are reduced with the wire of time. The territory is thus confronted to the risk of marginalisation, hollowness that can, as we will see below, make it lose its quality of 'territory'.

In the first part, we will demonstrate that the company is not an insignificant actor. We will remobilize cognitive approaches and more singularly the evolutionary approach, initially developed to include and understand the dynamics of companies to appreciate their territorial footprint.

In the second part, we will see the case of the Defence industry in the area of Brest, a singular territory as it has been formatted by this industry, which, since the beginning of the 1990s, has engaged deep reorganisations. Even if local stakeholders have proposed several alternative projects (offshore platforms, *Ship* dismantling, design and manufacture of windmills and tide-driven turbines) aiming at developing and valorising competences held by local actors, these latter still remain reticent today with regard to any diversification due to lock-in effects.

COMPANIES AND TERRITORIES

The Hidden Face of the Company

A company produces goods and services, but not only. There is also what is visible and what is not, at least on the short and medium term.

The company is not simply related to production, having for exclusive mission to combine factors of production (labour and capital) to reach a certain level of production. Therefore, it is not a simple organisation having for only target the reduction of costs of transaction, which would result from the imperfection of markets.

The company produces something else: immaterial and, in particular, specific competences (individual and collective), which will engage in the long term the territory where they are located in a technological trajectory and expose it to the risk of relative exclusion if this trajectory restricts its competences to a very specific field, excluding de facto alternative competences. The company cannot thus be only apprehended just in terms of optimal allocation of resources as done by neoclassic economists and in particular those who support a compromise approach (Williamson, 1994). The company must also be analysed within the framework of optimal creation of resources, a process that is not painless for concerned territories.

The company thus produces immaterial wealth that goes beyond classical inputs, namely standards, conventions, beliefs, competences, routines … as many resources from which the company builds its foundations, its difference and thus its advantage in terms of price and non-price competitiveness.

The company contributes to create labour or production norms, that is a way of producing goods and services. For instance, it can introduce an organisation of production, which structures behaviours of actors but also territories. By the goods and services it offers on the market and through its marketing activities, the company also takes part in the development of the standard of consumption, to which consumers refer and of which they become dependent … In that sense, the company can be defined as an institution (Thuderoz, 2010). It standardizes, formats behaviours and adapts its environment to its needs.

Beyond these norms, there are conventions, that is reciprocal systems of values governing the behaviours of actors that are produced through the interactions between actors. These conventions create order and thus ensure in the long term the production of physical and immaterial wealth. Moreover, among this immaterial production, there are individual and collective competences (e.g. tacit knowledge) that engage the company and the territory where it is located into a trajectory of development (Table 1).

Too much neglected for a long time by economists, this invisible part of productive activities forms the genetic legacy of the company. The latter builds its identity by the production of specific assets used by the company to differentiate but which, in the long term, generate irreversibility, inertias in its strategic choices. This situation has structuring impacts on the

Table 1. The Two Faces of a Company

Visible	– Goods – Services
Invisible	– Regulations – Agreements – Skills and competences – Routines

Table 2. An Approach in Terms of Capacity.

Production of competences

↓

Creation of specific credits

↓

Dependence of path

↓

Irreversibilities

↓

Reduction in the capacity to act

↓

Degradation of the attractivity of the territory

↓

Which public policy?

↓

To build capacities in the long term without freeing itself from its past territory, since the company's choices influence the evolution and opportunities of the involved territories. These latter, as underlined by the evolutionary school, are then engaged in a trajectory of development, at least on the short and medium term, thus resulting in irreversible choices and sunk investment in the strategies of actors and public policies (Ferru, 2009). Therefore, the company marks its seal on its territory (Table 2).

If all territories are confronted to irreversible choices, some of them know more important inertias, thus reducing the capacities of actors to be and act. The territorial footprint of the company can sometimes be so strong that the involved territory literally merges with the company.

THE COMPANY, ACTOR OF THE TERRITORY

A Multidimensional Footprint

While producing material and immaterial resources, the company marks the involved territory, that is those who live there and work there, in various manners. The footprint is:

- Industrial: revealed by the specialisation of products and activities related to the territory.
- Organisational: output norm (organisation of labour and mode of governance) and conventions that are specific to industry (decision-making based on shared beliefs).
- Sociocultural: for instance, the wait-and-see behaviour resulting from a very hierarchical and 'verticalised' organisation and a culture of secrecy like in the Defence industry.
- Psychological (generally unconscious): psychological traumatism that follows the restructuring or the closing-down of the company.
- Local policy: embedded influences on local economy and politics. For example, people in charge of local authorities are often former employees of these companies.
- Environmental: visual and spatial footprint (companies' plants, warehouses, infrastructures, industrial architecture, waste lands … but also protection of local ecosystems used as buffer zones).

If a change of trajectory is possible, it can only be considered as a long-term project as it is not spontaneous and can only be built by local stakeholders deeply implicated in the territory.

Path Dependence and Ability to Act

History Matters

The company generates tacit knowledge, which results from training and know-how acquired through repetition and experimentation. This knowledge is at the same time a source of competitive advantage and of inertias, and thus of dependence.

This production of specific resources goes through organisational routines, that is models of interaction that aim to provide solutions to problems encountered by the company. These routines are quite specific to a given firm and are not easily transferable and redeployed to other activities. The company can be distinguished by its competences, that is its tacit knowledge which is not codifiable and which, consequently, cannot be acquired on the market.

The company thus follows a path, determined by accumulated competences, resulting from the effects of training and experiences. Endogenous to its organisation, its specific assets determine the path of evolution, the trajectory of the company (Boissin, 1999).

Initiated by Nelson and Winter (1982), this evolutionary approach of companies' activities can be transposed to territories and thus help apprehend the factors of evolution or lock-up of territories. Marked by the strategies of companies (through past investment), which it shelters or had sheltered, a territory faces a dependence path.

In this approach, history matters (Dosi, Teece, & Winter, 1990). The past behaviours influence today's behaviours. Hysteresis effects are quite strong, and it is uneasy to depart from them. Problems of lock-in exist and remain in the medium term: a technological path can cause the insulation of a territory and result in its inability to adapt or evolve, thus leading to its impoverishment.

The situations of lock-in can be functional, cognitive (i.e. a specific vision of the world that limits the perception of opportunities or alternative options) as well as political (through institutional stalemate) (Laperche, Lorek, & Uzinidis, 2011, p. 346).

Ability to Be and Act

If the specific, material and immaterial assets at the grounds of competing advantage, they generate certain inertia, that create frozen features in a given territory, resulting in a reduction in the capacity of actors to act.

Borrowed from Sen (2000), the ability to be and act includes three main characteristics (Friboulet, 2010):

1. The ability is a freedom of choice, which supposes the existence of alternative possibilities and decisions being made in full independence. A gravitational territory is a territory that gives actors the ability to make decisions and thus enables them to control their fate.
2. The ability is a potential of blossoming. A gravitational territory creates diversity and supports the achievement of personal and professional projects.
3. The ability implies the means of achievement. It is indeed necessary that these possibilities of evolution or conversion can transform effective choices and eventually actions (Friboulet, 2010). The territory plays an essential part in the process of conversion of individual resources (financial, intellectual, social, human) into 'freedom to be and act' (Gérardin & Poirot, 2010, p. 30). It is thus necessary that facilitating institutions intervene to provide complementary resources (managerial support, financial support, training ...) that are required to achieve personal and professional projects.

BREST AND DEFENCE: MORE THAN 350 YEARS OF COMMON LIFE

Brest, the First French Naval Base on the Atlantic Ocean

Established at the edge of a deep river, La Penfeld, leading to a bay protected by a narrow maritime bottleneck, Brest enjoys an exceptional situation, which, for a very long time, represents a natural harbour.

Already a fortified place since the Roman Empire, Brest has really started its growth when Richelieu decided in 1631 to use it as one of the four naval harbours of the French kingdom. Since then, the life of the city is closely linked to the life of its naval base.

With Toulon on the Mediterranean Sea, Brest is today the main French naval base. In the 1960s, it became the home harbour of submarines that contributes to the naval component of the French deterrence. Therefore, Brest naval base has a strategic role in the Defence system of France.

This base, as most of naval bases, provides two missions: a service of Defence and an industrial mission. When adding sailors of the Fleet and civil workers of the base, there are today approximately 20,000 employees. Today, the industrial activity of this base focuses primarily on the maintenance and repair of surface vessels and submarines. It used to have dominant production activities, but they now represent less than 10% of turnover. This industrial activity is delegated primarily to DCNS, a former arsenal that was transformed into a public company in 2000. DCNS and its subcontractors gather today a workforce of approximately 3,500 people.

Following the several restructuring of Defence activities since the early 1990s, the naval base has lost its importance even though it remains by far the most important economic pole in the area of Brest. At the beginning of the 1990s, whereas the reorganisations were not yet really committed, Le Nouail, de Penanros, and Sauvin (1995) had calculated that, in terms of direct, indirect and induced employment, approximately one-third of Brest households were economically dependent on the activities of the naval base. Today, the ratio has undoubtedly fallen, but it seems that 20% or 25% of households are dependent, which remains extremely important.

Territorial Footprints of the Navy

Three hundred and fifty years of presence on a territory inevitably leave traces.

The territorial footprint of the Navy is multiform. It marks at the same time the physiognomy of the city, its economic and political life, the mentality and the way of life of its inhabitants.

Impacts on Town Planning

Brest is a city divided into two parts. Its natural central axis, La Penfeld River and its banks, constitutes an enclave of the State within the city. It is a closed space, entirely dedicated to the activities of Defence, which remains out of the jurisdiction of the city. Only servicemen and the inhabitants of Brest who work inside the arsenal have access. The military specialisation of Brest is therefore spatially perceptible and leads to two separate spaces that have limited interactions.

Brest is a historical city but without ancient buildings. Because of its position of naval harbour, Brest was completely destroyed during the Second World War. Rebuilt in a hurry after the war, the new city did not keep traces of its past. Only some paintings and engravings of Navy painters or, for the more recent period, some postcards give inhabitants of Brest an idea of what Brest was before. This reconstruction has favoured the footprint of the Navy on Brest, since it primarily served to improve the effectiveness of activities related to the Navy and its industrial partners.

Brest is a city that misses bridges. Taking into account the constraints posed by the Navy (a bridge with mobile apron to allow its ships to pass below without having to dismount their antennas), the city gave up the construction of a bridge that would have doubled the bridge raising at the entry of La Penfeld River. Such a bridge would have allowed a better and denser circulation between the two banks of La Penfeld River. Additionally, for strategic reasons (to avoid any risk of obstacle in the circulation of its submarines), the project of bridge spanning the narrow part of the bay has never been accomplished. Such bridge would have allowed a fast connection between Brest and its hinterland in the southern bank.

Sociological and Political Footprints

Footprint on trade unions. Until 2000, industrial activities were mainly conducted by a State arsenal, *Direction des Constructions Navals*, and companies, which were at least partially owned by the State. This led to a strong influence of trade unions, and the State had to be exemplary in the application of labour regulations within its own sites. Employees acquired for ages a right of expression and a strong representation of their trade unions in Defence sites. This historical setting, going back at least to the 1940s, favours a culture of commitment and the creation of strong and influential local trade unions.

Footprint on many sport and socio-educational structures. One point is quite surprising in Brest. There are several sports and socio-educational associations that rely on many employees of the naval base to carry on activities and administrative duties on a purely voluntary basis. This strong implication can undoubtedly be explained by.

- An encouragement via the action of trade unions to take responsibilities
- A need to express people's social utility differently as through the production or the maintenance of arms systems
- The possibility for State employees to have free time given by the armed forces in order to undertake these external activities.

A strong representation of employees of the naval base in the city council. Partly undoubtedly for the reasons that one has just mentioned, employees or retired people of the naval base are present in the town councils of Brest and surrounding cities. As in Toulon or other Defence-oriented places, people from the realm of Defence are overrepresented in the city council.

Cultural Footprint

It acts initially of the culture of company maintained for a long time by the naval industrial plant and which still deeply marks the working behaviours and industrial relationships. Three features mainly concern this culture:

1. the quality of the products carried out. It concerns the safety of the soldiers who will have to use them. The culture of the beautiful work, of the time to devote to it that cannot be counted, is always strongly present within the workers of the arsenal, even if the requirements of profitability which are essential today even within the States services attenuate it from now on the range of it.
2. the bulk-heading of the activities and
3. the culture of the secrecy.

These two last features are narrowly bound and strongly mark the activities of Defence. They appear by a tendency to the constitution of own suppliers and subcontractors networks, the limitation of the exchanges with the external environment, and a reserve to communicate at the point to manage from there to sometimes develop same research within services of the same establishment!

The cultural print is still more largely this inclination of Brest and its elected officials to always wait for the feeder State, and initially via the budget of Defence, a response to all the local difficulties. Undoubtedly a consequence of the professional origin of several of them, it is finally this tendency of the municipal elected officials to adopt the point of view of the Navy, even if it goes against the interests of the city.

Economic Footprint

The installation by Richelieu of the naval base marked the Brest-native economic landscape definitively. While the Navy base takes a large place in the city, the economic dimension of naval activities results mainly from DCNS, the naval shipyard. This latter was set up at the same time as the Navy base, and it has developed steadily with the French Navy.

Brest is one of the very first French naval industrial poles today. DCNS is the first private employer and it has several companies relying on its own activities, including a large number of SMEs. Therefore, DCNS represents the heart of the local industrial system, with a strong economic footprint. This is reinforced by the fact that there is no alternative industry able to match naval activities in the area of Brest.

Therefore, when one combines activities related to the Navy and activities engendered or steered by DCNS, it is obvious that the economy of Brest is heavily structured around Defence-related activities.

Obstacles and Opportunities for the Development of the Local Economy

In Brest, the Navy contributes to important missions for the international security of France, particularly through the naval component of the French deterrence. Therefore, these missions appear to be more important than local economic interests. This priority given to the military function of Brest generates particular constraints for the development of this territory. This is particularly true concerning the choice of Brest naval base as the sanctuary of the naval force of nuclear dissuasion.

However, obstacles that are specific to Defence activities do not eliminate any opportunity for the economic development of Brest area. They only determine the paths that this territory is allowed to follow.

Despite these limitations, the presence of the Navy opens opportunities of local development.

Regarding tourism: The presence of the Navy in Brest, especially its castle with its historical collection, its naval harbour and its warships, constitutes a great opportunity to develop touristic activities. Nevertheless, this potential is limited by the lack of appropriate infrastructure, notably the absence of easy and fast connection with the left bank of the bay, which would be possible if there were a bridge on the bay or a ferry. This potential development is hindered since Brest appears in fact to be in a situation of tourist 'cul-de-sac'.

Regarding urban development: The progressive withdrawal of the Navy from its old bases along La Penfeld River as well as the renunciation of Defence castles, batteries and various sites of observation spread out along the bay open opportunities for concrete projects of urban development. One could imagine the installation of new districts, the constitution of new commercial harbour infrastructures, the opening of new leisure spaces, etc.

Nevertheless, the true urban opportunity would come from the retrocession to the city of the whole enclave of La Penfeld, a space today underoccupied by the Navy but that does no longer have a true military utility. By offering a new central district to the city, such a retrocession would result in a new urban dynamism, which could lead to a complete remodelling of the city.

Leveraging on the industrial competencies: Due to the naval shipyards and Navy-related activities, the area of Brest constitutes an industrial pole gathering many skills and know-how. Indeed, this industrial pole requires advanced technology, a state-of-art know-how and skilled workers able to manage technological challenges.[1] This represents a real asset for Brest. Today, two opportunities are offered to Brest to develop this asset that fit perfectly in the continuity of its trades and its knowledge to make naval.

The Dismantling of Ships

Historically, the Navy used to get rid of its ships by sinking them. However, regulations have changed and the Navy has to manage the environmental impacts of its fleet, especially in the end-of-service phase. Problems arising from asbestos and new preoccupations with the protection of ecosystems have resulted in the definition of strict standards when the dismantling of old ships is concerned.

These requirements open an opportunity for Brest and its industrial pole to redeploy its activities from declining naval market towards a new business. This kind of conversion should be facilitated by the commonalities between both activities.

For such a transformation, it is necessary to overcome the reserves of the naval sector. In fact, this latter considers that dismantling activities are not as gratifying as the production of ships, that they consider as the most noble activity, even compared to repair activities, which they place between the two kinds of business (Fig. 1, Panel A and B).

Developing such an activity requires that local stakeholders change their perspective and integrate dismantling as the natural step in the lifecycle of ships. In a sustainable development perspective, construction, repair, dismantling/breaking become the three phases of the production process, which are interconnected, each one preparing the following and being also essential to the reproduction of the process. There is particularly a recycling dimension to take into account here.

It is necessary that naval workers give up their discontinuous and hierarchical approach of naval activities and accept a continuous and more interdependent approach based on a three-phase process. Without such

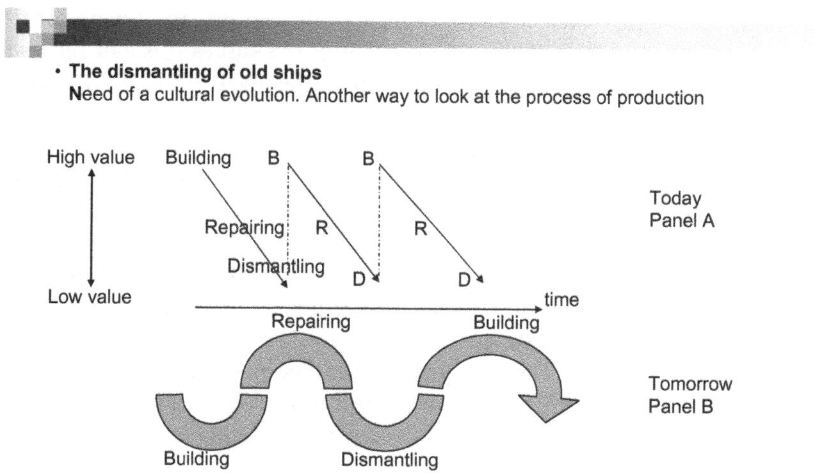

Fig. 1. Building, Repairing and Dismantling, Three Phases of the Production Process.

a change of mind, it is likely that dismantling activities cannot develop in Brest and this market opportunity is bound to lose for the benefits of other territories. This was already the case for the French aircraft carrier *Clemenceau.*[2]

The Construction of Marine Wind Mills and Tide Turbines

The energy transition, which leads to the development of renewable energies, offers good opportunities for the territory of Brest because of its strong competencies in naval and maritime systems.

The winds that sweep its coasts, the currents, which brew the seas that surround Brest, make Brittany an area predestined for the installation of offshore wind fields and tide turbines.

The realisation of these complex machines presents many similarities with shipbuilding and requires advanced technologies that are shared with naval systems. Brest could become one of the industrial poles in the field of maritime renewable energy. Brest has considerable competences but also specific assets in its naval and industrial areas available to develop and assemble such systems; and the city benefits from a harbour infrastructure to transport these systems.

Such an evolution nevertheless requires a support from the leading company in the territory, that is DCNS. This constitutes a possible bottleneck, since this company has always been careful about its diversification towards civilian activities. Taking into account the stakes this can represent in terms of employment for the Brest area, the local community has to show sufficient determination to push DCNS towards these new activities and leverage on its numerous competences and assets.

The activities of Defence are not by themselves a decisive obstacle vis-à-vis the development of the territory. On the contrary, they even sometimes offer important opportunities to nurture such an evolution beyond Navy-related activities. The principal obstacles to such an economic evolution are initially of cultural nature: obsolete prejudices against some activities (military vs. civilian activities, construction vs. dismantling) and a tendency of the local community to align on the positions of the Navy, demonstrating that after more than 350 years of common life, the emancipation of the city owing its existence to the Navy is not yet completely accomplished.

CONCLUSION: FROM SPECIALISATION TO SPECIFICATION

When one analyse the case of Defence-dependent cities, it appears that there is not one territory but at least two territories of different history and culture. Some of these cities, more than others, miss plasticity. It is the case of the area of Brest, a territory marked by a strong degree of specialisation in terms of activities and competences. The core activities (industries of Defence), which were developed there, are characterised by a very strong impact with regard to other activities located on the territory. This juxtaposition of activities results from the specificity of the organisational structure of the Defence industry, favouring a very hierarchical and vertically structured organisation that is unfavourable to technological externalities and spinoffs in non-Defence activities.

Regarding this problem of technological and organisational lock-in, it is important to consider another mode of development, based on a process of specification, the objective being (following the identification of competences held by local stakeholders) to develop them in other fields. Such dynamics would result in a coherent diversification with the competences built locally by companies as well as workers.

However, the passage from specialisation to specification is not easy (Colletis & Gilly, 1999). During the period of specialisation, the territory builds up specific resources and assets (dedicated to a given activity) that cannot be redeployed or, at least, not easily redeployed towards other activities.

In the same way, the sociocultural, psychological and political footprint of Defence constitutes a powerful barrier to a possible change of trajectory of development. The redeployment of competences supposes that there is an organisational flexibility of the economic fabric. It is necessary also that there is a strong density of relations between actors and stakeholders. These more horizontal relations must be grounded on a minimal confidence and actors must support new productive solutions. They have to be conscious that the territory shelters all competences required to carry out their projects. The process of specification cannot thus start without the preliminary existence of an institutional resource. This latter is essential to coordinate stakeholders and organise the redeployment of the territory beyond its initial specialisation in order to reveal and activate the resources held by the actors.

If the Defence footprint can constitute a handicap, due to irreversibility and sunk investment in the short and medium term, it can also represent an advantage by creating opportunities. A marked territory is a territory that is different from others and that knows where to go. The footprint creates a path to follow, reducing the uncertainties when lacking choices for the future of the territory.

It is thus necessary to lay down a public policy whose objective is to build capacities to act in order to transform local stakeholders into actors. This is the necessary condition in order that the territory remains a territory, because a territory without project is a territory which dies and which is likely to become a simple geographical entity. To consider a rebuilding of the cognitive trajectory of the territory, rebuilding which passes, beforehand, by a work of deconstruction and not of demolition. Here, the intervention of the authorities (particularly, local authorities) and the civil society are essential.

NOTES

1. In the late 1990s, naval workers succeeded in building two offshore platforms in only eighteen months. They proved as such their ability of adaptation to new activities.
2. After many adventures and several months last on standby of dismantling to the wearing of Brest, fault of local will, it is with an English building site that will be entrusted the dismantling of this old aircraft carrier.

REFERENCES

Boissin, O. (1999). La construction des actifs spécifiques: Une analyse critique de la théorie des coûts de transaction. *Revue d'Economie Industrielle, 90.*

Colletis, G., & Gilly, J.-P. (1999). Construction territoriale et dynamiques économiques. *Sciences de la Société, 48.*

Dosi, G., Teece, D., & Winter, S. (1990). Les frontières des entreprises: Vers une théorie de la cohérence de la grande entreprise. *Revue d'Economie Industrielle, 51.*

Friboulet, J.-J. (2010). La construction de l'attractivité: Une analyse en termes de capacité. *Mondes en Développement, 149.*

Gérardin, H., & Poirot, J. (2010). L'attractivité des territoires: Un concept multidimensionnel. *Mondes en Développement, 145.*

Laperche, B., Lorek, M., & Uzinidis, D. (2011). Crise et reconversion des milieux industrialo-portuaires: Dépendance de sentier ou renouveau économique? Les exemples de Dunkerque (France) et de Gdansk (Pologne). *Revue d'Economie Régionale & Urbaine, 2.*

Le Nouail, M. N., de Penanros, R., & Sauvin, T. (1995). Activités militaires et expériences de diversification dans la région brestoise. In Reconversion des industries d'armement sous la dir de Roland de Penanros, La Documentation Française.

Nelson, R., & Winter, S. G. (1982). An evolutionary theory of economic change. Cambridge, MA: Belknap Press/Harvard University Press.

Raveyre, M. (2005). Introduction au numéro spécial restructurations. Nouveaux enjeux. *La Revue de l'Ires, 47.*

Sen, A. (2000). Un nouveau modèle économique. Développement, justice, liberté, Odile Jacob.

Thuderoz, C. (2010). *Sociologie des entreprises* (Collection Repères). Paris: La Découverte.

Williamson, O. (1994). *Les institutions de l'économie* (trad. Française). Paris: InterEditions.

FURTHER READING

Boschma, R. (2004). Proximité et innovation. *Economie Rurale, 280.*

Bouba-Olga, O. (2003). L'économie de l'entreprise, Points, Seuil.

Charles, E., & Sauvin, T. (2011). Quand un label révèle un autre modèle de développement. Le cas du label 'Accueil Pêche Finistère'. Revue Téoros, à paraître.

Colletis, G. (2010). Co-évolution des territoires: Une perspective institutionnaliste. *Revue d'Economie Régionale & Urbaine, 2.*

Coriat, B., & Weinstein, O. (1995). Les nouvelles théories de l'entreprise. Le livre de poche.

Ferru, M. (2009). La trajectoire cognitive des territoires: Le cas du bassin industriel de Châtellerault. *Revue d'Economie Régionale et Urbaine, 5.*

Lajarge, R., & Roux, E. (2007). Ressource, projet, territoire: Le travail continu des intentionnalités. In H. Gumuchian & B. Pecqueur (Eds.), *La ressource territoriale.* Paris: Economica.

Léon, A., & Sauvin, T. (2010). L'entreprise et son empreinte territoriale: Quelle politique d'attractivité? *Mondes en Développement, 145.*

Zimmermann, J.-B. (2005). Entreprises et territoires: Entre nomadisme et ancrage territorial. *La Revue de l'Ires, 47.*

LOOKING AT MILITARY LANDSCAPES: DEFINITIONS AND APPROACHES

Rachel Woodward

ABSTRACT

This chapter explores how military landscapes have been conceptualised and understood. The chapter starts by defining what is meant by the terms 'landscape' and 'military'. The chapter then proceeds with an exploration of a range of examples from a variety of disciplinary origins in order to support the argument that military landscapes constitute a diversity of sites and have a ubiquity of occurrence. Such examples include battlefields and other sites of conflict, the interconnections between landscapes and the pursuit of specific campaigns and conflicts, the issue of environmental impacts of military activities and the interpretation of these with reference to the specificity of landscapes, and landscapes of memory and military memorialization. The chapter then goes on to consider how military landscapes can be viewed, raising questions about the visibility and invisibility of such sites. The chapter concludes with some observations about the imperative for sustained scholarly attention to military landscapes, in order to inform debates about militarism as a social force.

The Evolving Boundaries of Defence: An Assessment of Recent Shifts in Defence Activities
Contributions to Conflict Management, Peace Economics and Development, Volume 23, 139–153

ISSN: 1572-8323/doi:10.1108/S1572-832320140000023010

This chapter is about military landscapes and how they might be defined and studied. It is predicated on a concept of landscape that is inclusive of a number of different approaches to the term, and on an understanding of militarism and military activities as far-reaching in terms of how they shape economic, social, cultural and political life across the globe. In this chapter, I review a selection of studies of military landscapes, studies which may or may not have been undertaken with an explicit orientation towards the concept, but which are included here to illustrate the range of ways in which military landscapes can be understood. The variety here extends from explorations of the morphology and interpretation of battlefields, through to accounts of the meanings attached to sites as diverse as military barracks and military memorials. The chapter then considers some issues pertaining to the investigation of military landscapes, particularly those concerning the visibility and invisibility of these sites. The intention here is not to provide a definitive overview of this diverse field, but to bring together diverse literatures to demonstrate how military landscapes can be interpreted from a range of perspectives.

In using the term 'landscape' I refer to three related ideas. The first of these is the idea of landscape in morphological terms. Through this definition, we understand landscapes as the material patterning of land brought about through processes that are both natural and human in origin. In referring to landscapes as a focus for study, in the first instance this includes the study of the ways in which features and objects appear in space, and the observation of the distribution of such features and objects (Muir, 1999). The second conceptualisation of landscape is that of the landscape as text (Cosgrove & Daniels, 1988). Long-standing as an approach in much cultural geography, this conceptualisation promotes the reading of a landscape as one would read a text, with a view to establishing the social, economic and political relations which cause such landscapes to come into being, and through which we can ascertain the consequences of the establishment of visible manifestations of such processes. When we talk of the representational capacity of landscapes, the ways in which landscapes might be read in terms of how they speak to specific ideas, it is to this conceptualisation of landscapes that we refer. The third idea referred to by the term is of landscapes as sites of experience, locations through which and in which subjectivities are constructed and articulated, which evince emotional and affective responses and which, in turn, are read in ways that are shaped by the responses provoked (Wylie, 2007). In this chapter, I use all three of these ideas about landscapes in order to explore how military landscapes can be understood and studied.

Talking about military landscapes requires us also to define the term 'military'. In this chapter, I refer primarily to landscapes that are military in origin, which is to say that I refer to landscapes which are produced, read and responded to as an outcome of the organised activities of those institutions (armies, navies, air forces, defence ministries) which have been vested by the state with the authority to exercise potentially lethal force and violence. Military landscapes therefore do not, primarily, refer to landscapes where the imprint of security agendas by non-state actors can be seen, and nor do they refer primarily to landscapes which see the impact of paramilitary activity. In this chapter, I also use two terms quite explicitly. One of these, 'military activities', refers to the activities undertaken by military forces and state defence organisations in both conflict and non-conflict situations, either in direct pursuit of military objectives, or as a by-product or secondary consequence of the pursuit of military objectives. Military activities are thus not just the explicit and overt use of lethal force in combat operations, but also the vast range of planning, preparatory and training activities without which the execution of military force would not be possible and through which realms of defence are constructed and sustained. The other term used is 'militarism' which refers to the extension of military objectives and ideas into non-military realms of social life. Whilst the term may sometimes be used to denote quite extreme examples in historical periods where this process has been understood as quite explicit, my argument here is that militarism is more usefully used as a concept with which to understand the logic and consequences of more mundane, prosaic and everyday experiences. Furthermore, this allows us to see militarism as it operates at a range of scales, from the national right down to that of the individual (see Woodward, 2005).

In this chapter, I argue that military landscapes are, in advanced capitalist economies, ubiquitous. Military landscapes are everywhere. If we take together the three observations made about how we can understand landscapes, and the observation about the need to focus on both military activities and militarism, it becomes possible to start looking at the wide range and variety of sites and spaces which we might denote as military landscapes. I will return to this point in the conclusion, in order to argue for the necessity of sustained academic inquiry into such landscapes and address the question as to why we should look at these things. In the first half of this chapter, though, I want to explore something of this ubiquity. I also want to underscore something of their subtlety. A vignette from my academic life will illustrate this point. One afternoon, a couple of years ago now, I was waiting for a train at Newtown railway station in rural Wales and strolling on the platform with colleagues at the start of the journey home after a conference. I happened to glance at one of a number of doors in a building on the platform. On it, at waist-height, a small sign read 'Thales', lettered according to a distinctive corporate visual imagery. The Thales Group is a multinational corporation, French in origin, specialising in information and communications systems for the defence sector. So here was a corporate advertisement for the military-industrial complex in a small railway station. Thales at the time was part of a consortium supplying communications technologies under contract to Network Rail, the company which runs Britain's rail infrastructure. Thales is representative of the ever-increasing economic and political power of private corporations, the core purpose of which is assisting with the planning, provisioning and prosecution of armed violence under the neo-liberal regimes which shape our militarised geo-economic present (see Cowan & Smith, 2009). The Thales sign at Newtown railway station may not militarise the landscape in ways that are as obvious as a bomb crater or a tank, but that glance at that door confirmed for me something about our need, as researchers of military phenomena, to grasp both the subtlety and ubiquity of military power in shaping the landscapes which surround us, as well as its more obvious, immediate and specific manifestations.

THE VARIETY OF MILITARY LANDSCAPES

The most immediately obvious type of landscape which we would recognised as military would be sites which bear the distinct imprint of military action and armed conflict. We often term such sites 'battlefields', although in fact the range of types of sites included here escapes the idea of a single defined rural space involving the direct engagement of opposed, organised armed forces, and would include urban areas as much as it would the more traditionally understood sites of battle such as are still so evident in contemporary Europe. The study of battlefields constitutes a major component of both academic and lay interest in military landscapes. In traditional military geography, for example, the exploration of the interplay between physical geographical and human factors and the consequences of this for the outcome of specific historic battles and campaigns constitutes a long-standing interest (see Doyle & Bennett, 2002; Palka, Galgano, & Corson, 2005). Examples of specific studies within this approach include Passmore and Harrison's (2008) exploration of the Battle of the Bulge in the

Ardennes, Belgium, in 1944, or Pollard's (2009) reading of the field of the battle of Culloden, fought between Jacobite and Hanoverian forces in 1746 in Scotland, with reference to the terrain on which it was fought and its influence on the battle's outcome.

Viewing military landscapes through a wider lens, we can also consider here the work of environmental historians and the contributions their scholarship makes in providing interpretation of campaigns and conflicts with reference to physical and human geography at the larger scale (see Pearson, 2012; Pearson, Coates, & Cole, 2010). Examples would include Pearson's research on Vichy France and the role of landscape, discursively and materially, in encounters between German forces and French resistance in the Camargue (Pearson, 2009), on the unoccupied zone of France in *maquis* areas (Pearson, 2008), and on the role of forestry and foresters (Pearson, 2006, 2007). Although not working within an explicitly environmental historical framework, the work of Peluso and Vandergeest (2011) is complementary here in its explorations of 'political forests' in South East Asia and their material and discursive uses during the regional conflicts precipitated by the Cold War.

Military landscapes can also be considered as sites in which environmental issues and politics are played out in the present. There is long debate within geographical and environmental studies about the consequences and politics of military environmental impacts. Military activities, ranging from active operations through to peacetime training and maintenance activities, are often seen as inherently environmentally destructive (see Woodward, 2004). Examples include the contamination of soils, air, water, flora and fauna through the use of explosives, heavy metals in munitions and other chemical pollution, and geomorphological changes as a consequence of bombing, shelling or deliberate earthworks and environmental manipulation which may be environmentally deleterious. Contamination as a result of training, transportation and maintenance activities is also an issue. The military forces of many nation-states, particularly those of advanced capitalist economies, have increasingly been held to account by national governments and by pressure groups and the environmental lobby over environmental impacts, and it is worth noting how external critique has often prompted remedial action. However, and as a number of commentators have noted, the levels of environmental destruction and contamination produced as a consequence of military activities have been of sufficient severity or extent as to merit explanatory strategies and explanations rife with political intent concerning armed forces' self-image (see Davis, 2005, 2007; Davis, Hayes-Conroy, & Jones, 2007; Havlick, 2007, 2011). Critiques have moved beyond the simple idea of a 'greenwash' to focus on the complex interplay between defence objectives around the use of landscapes, and defence objectives around the presentation for public consumption of explanations around the uses of such landscapes.

We can also consider military landscapes as sites encountered by both military personnel and civilians alike in the course of daily lives in contexts far removed from armed conflict. Military installations such as bases, dockyards and airfields, communications complexes, barracks, training areas and depots occupy significant amounts of land in many nation-states. An often-quoted estimate by Westing (1988) suggests that at least 1% of the land area of the major industrial powers is used for military purposes. Clearly, if we are looking for reasons to explain the ubiquity of military landscapes, the sheer number of installations and their wide distribution would be key here. Such installations provoke reactions. Landscapes, as sites which are viewed, inhabited and experienced, exert effects on individual subjectivity and emotion. Tivers' account of the landscapes of a major British Army basing complex, Aldershot, in the south of England, explores the different dimensions of such responses by looking at axes of experience. Such landscapes may be simultaneously places of stress and security, of stimulus and ennui, and of status and stigma (Tivers, 1999). Whilst studies of military basing, whether home or overseas, may focus primarily on the politics of such installations, in landscape terms there is much to be learnt from a consideration of the individual, personal responses these landscapes provoke both amongst civilians who live in proximity, but also amongst the personnel who live or work within them. Accounts of such landscapes and the complexities of the politics they provoke include Higate and Henry's (2011) exploration of the zones of division in Cyprus maintained by UN personnel, or Yamazaki's (2011) exploration of the militarisation of space by the US armed forces in Okinawa, Japan.

It is in contexts of active armed conflict that we see most clearly the imprint of military power across landscapes. Studies of the landscapes of conflict are, because of the changing nature of warfare, focusing increasingly on urban sites where 'war amongst the people' (Smith, 2006) is played out. A key line of inquiry here concerns what Graham (2009, 2010, 2012) terms the 'new military urbanism', which examines the ways in which military imperatives for the domination of space have found new expression due to the demands imposed by the urban context in which active operations and less visible securitisation strategies take place. The landscape consequences of urban operations are often the starkest of the military landscapes of armed conflict, given that they shape the very signs and symbols of modern development. These landscapes are utterly different to the battlefields of the 19th century, as Weizman's (2007) exploration of the Israeli armed forces' complex architecture of occupation in the urban areas of Israel/Palestine makes clear.

Military landscapes can also be seen as sites where memorialisation of military activities takes place. Although not necessarily immediately evident as locations bearing the imprint of military power in place, in the aftermath of conflict (and quite apart from visible remains of conflict) we can see how militarism extends to physical form, and becomes enmeshed in processes of memorialisation at scales from the national to the personal. Acts of preservation of battlefield sites are a significant example of this process in action. Former battlefields can become places for the construction and re-telling of narratives of national identity. The example of the War in the Pacific national historic park on Guam is a case in point, where the landscape of the park invites a reading of Guam which supports a very specific war memory of Guam as a US colony and base, and as a significant site for US forces during the Second World War (Herman, 2008). With reference to Hawaii, Ferguson and Turnbull's (1998) reading of that island focuses on the imprinting across the landscapes of the island the power of US military forces and the island's construction as a significant outpost in support of US military territorial objectives. The meanings of battlefields, and of the conflicts which produce them, are fluid and unstable, subject to changing social understandings and geopolitical contexts. So, for example, studies of the Culloden battlefield (Masson & Harden, 2009; Pollard, 2009), the Isandlwana site of the battle between British and Zulu forces in South Africa in 1879 (Pollard, 2007) and sites in Delhi involved with the Indian revolt or first war of independence in 1857 (Lahiri, 2003) all show how the meanings and interpretations of such battlefield sites shift over time in response to changing social responses to the conflicts which they remember, and the specific activities that occurred there.

Military landscapes can also be sites of mourning and remembrance, whilst they also simultaneously promote interpretations of specific battles, campaigns and conflicts. The landscapes of the First World War, particularly those of the Western Front, have long intrigued scholars for the way in which the necessity of assertions of identity at national, regional and regimental scales intertwine with the necessity for the remembrance and memorialisation of the war dead as individuals. It is also possible that the scale of scholarly interest in such sites reflects the fact that the practices of memorialisation on the Western Front constituted such an extraordinary period of symbolic landscape creation in the early 20th century (Bushaway, 1992; Heffernan, 1995). Site-specific assessments of the politics and practices of national and individual memorialisation include studies of the Newfoundland memorial at Beaumont Hammel on the Somme (Gough, 2004), the Canadian national memory at Vimy ridge (Hucker, 2009), the South African memorial at Delville Wood (Foster, 2004), the Passchendaele site on the Somme (Iles, 2003), the Ulster Memorial Tower on the Somme (Switzer & Graham, 2010), and the British war cemeteries and the ideas mobilised through their design (Morris, 1997). We see in such readings of landscape how the morphology of specific landscapes provides a form of text to be read in order to understand prevailing political ideas and sentiments both during a memorial site's construction and subsequently.

When considering how military memorial landscapes work as places of remembering, we can also consider how the political imperatives and personal needs of those involved in their construction and maintenance, and of those who live and visit, change over time. The significant body of work on the memorial landscapes of the Second World War in Singapore (Muzaini, 2006; Muzaini & Yeoh, 2005a, 2005b, 2005c, 2007) shows how the readings of memorial landscapes (including museums) demanded of the post-colonial nation, with an emphasis on national and humanitarian readings, have replaced older interpretations that spoke to both individual and personal remembrance, and to assertions of imperial power and memory. Similarly, as Dimitrova (2005) has argued with reference to Bulgaria, individual and local modes of remembering of the Bulgarian dead of the First World War were reworked and developed over time in response to a developing state articulation of this conflict in nationalistic and patriotic terms. Subsequent generational re-interpretations of Second World War dead in Finland, with a more contemporary emphasis on individual memory, overlay older state-sanctioned modes of memorialisation with an emphasis on national identity, established in the immediate aftermath of the conflict by the Finnish state (Raivo, 2004). It is also pertinent to remember that even in the process of their construction, there can be quite distinct differences as to how these memorial landscapes take the form that they do and thus promote certain readings and responses over others. For example, the Broken Hill First World War memorial in this Australian mining town could be read as a snub to the resistance to the war of the unionised labour force by the town's political elites keen to respond to the call to arms made by Britain as the imperial mother country (Rainbird, 2003).

VIEWING MILITARY LANDSCAPES

Military landscapes, then, comprise a diverse array of types of sites. An interesting question then follows, as to how we can *know* these sites. There are two elements to this question. The first of these concerns the visibility or otherwise of military landscapes, for whilst claiming their ubiquity and wide distribution, we should also be aware that some are more visible than others, and there are reasons for and consequences of this general visibility which have knock-on effects on the extent of our knowledge of military landscapes. The second element to this question concerns how we might look at or encounter these sites. This issue goes beyond the simple fact of visibility/invisibility, and considers the methods which we can deploy to establish understanding of the politics and power of these places. In the following section, I consider these two elements of the question of knowing in turn.

It is a truism that some military landscapes will be more visible than others. Localities which are used by military forces for military purposes tend, as a rule, to be subject to security controls which render them more or less visible to passing (or even sustained) civilian interest. This may be true for sites of active operations as well as for the more prosaic and everyday sites of domestic military basing. These are sites where sight is restricted, in terms of observation of the internal operations contained within. So, the voluminous literature on military landscapes which focuses on landscapes of memory and memorialisation may, perhaps, be attributable at least in part to the ready access of civilians to such sites, located as they are in public space and constructed with the express intent of prompting civilian access. The relative absence of sustained studies of military landscapes which are locations of military operations and preparations may reflect the limitations on public access even to sites of very low or negligible security value. Such sites, as Tivers (1999) and Woodward (2004) make clear, can be intimidating to non-military visitors.

The restricted opportunities for viewing and access are not insurmountable, however. In my own teaching, for example, which includes a module where students are encouraged to conduct personal fieldwork and write an account of the geographies of a military installation, there will be examples every year, where an enterprising group or individual has formally requested access to a military installation or base for a briefing on its function and purpose, and this has been granted. We cannot assume, necessarily, that all sites are always off-limits. But many of course are. The work of Trevor Paglen as a geographer interested in these secret sites is instructive here. He details, for example, the lengths to which he and companions have gone to trek through the deserts of Nevada to observe the hidden, secret spaces of military testing in the American south-west (Paglen, 2009).

But there are further elements to this question of visibility. One element concerns public knowledge beyond that which can be gleaned through in-place observation. Are sites mapped? Do they appear on cartographic representations of space produced for civilian purposes? In the United Kingdom, the Ordnance Survey map series (itself of 18th century military origin) for much of the post-Second World War period famously contained absences – white spaces (literally) where existing military installations thought by government and military forces to be of a sensitive or secret nature were simply not marked. Although visible on the ground, in cartographic terms such sites were rendered invisible. With the development of orbiting satellites for both military and civilian purposes, and the development of technologies for the high-level photographic capture of the earth's surface, and with the expansion into the public domain of such photographic representations through facilities such as Google Earth and StreetView, such cartographic omissions have become increasingly irrelevant. This is not to say, however, that the representations produced by Google Earth and suchlike offer a ready glimpse of a contemporary reality, for such images are of course famously subject to editing and control in terms of what is shown (Perkins & Dodge, 2009). The point here is that we cannot assume that military landscapes are necessarily knowable, or unknowable – the point, in terms of empirical exploration, is to ascertain visibility and invisibility as a necessary part of investigation. We should also consider the wider, more abstract issue of what it means to see. The focus of much contemporary critical geopolitical inquiry has been precisely this issue of visualisation and the scopic regimes which shape how we understand what we see (see MacDonald, Hughes, & Dodds, 2010). These are pertinent issues for the theorisation of military landscapes.

CONCLUSION: THE FUTURE NECESSITY OF STUDYING MILITARY LANDSCAPES

Ultimately, there is also a question as to why we should look at military landscapes, and about the questions we should ask of these spaces and places in view of contemporary changes to the organisation and structure of military forces and institutions. For what reasons is there scholarly benefit to the investigation and viewing of such sites? I suggest that there are three specific arguments to be made. The first of these is about visibility. The process of looking or attempting to look at military landscapes is a process which renders phenomena visible. At its most simple, this entails taking the ordinary, the prosaic and the taken-for-granted and subjecting them to critical gaze. An example here would be Tivers' work (1999) which takes the simple phenomenon of a town dominated by a complex of army facilities, and builds up a complex argument about the range of social responses that mundane sights such as signage and buildings provoke.

The second of these is about accountability. The investigation of military phenomena is an investigation of the accountability of military forces. Whilst this relationship may be more explicit in analysis of issues such as defence expenditure or military personnel strategies, where clear lines of public expenditure can be accounted for or questioned by analysis of the outcomes of such expenditure, the question of military accountability to the civil society on behalf of which it operates is pertinent too to landscape issues. An example here would be the ways in which memorial landscapes, which are so often state-constructed with direct input from military organisations, may (or may not) accord with civilian desires to see the costs of armed conflicts recognised in ways which make such conflicts meaningful to them.

The third argument for sustained critical investigation of military landscapes concerns militarism, and the academic imperative to understand how, and under what circumstances, the extension of military objectives to civilian spheres of social activity occurs. Military landscapes are spaces and places where the marks of military imperatives are manifest in material form, and which can be read in order to understand the workings of such imperatives. By understanding how militarism operates as a social force across space, we develop the basis for better-informed social debates about the extent to which civil society may wish to express its support, rejection, acquiescence or denial of the military activities undertaken in its name.

These questions of visibility, accountability and civil-military relations are highly pertinent when we consider military activities and militarism not as enduring and monolithic over time, but rather as dynamic, shifting processes. Changing geopolitical configurations, combined with socio-economic and political alterations and differentiations internal to nation-states, continue to effect change in the structure, organisation and thus spatial distribution and entrenchment of military forces and activities. One example here would be the changes seen in contemporary European nation-states since the end of the Cold War, in terms of the significant reduction in sizes of standing armed forces, the movement in many nation-states from a conscripted to an all-volunteer force, and evolutions in the use and capabilities of weapons systems and doctrines of warfare. These are changes which take place – literally – and the consequences of such changes are written upon the landscape. The research agenda suggested by contemporary changes is in part about understanding and thus predicting consequences for spaces and places as a consequence. The return to civilian ownership of redundant military lands would be one example, and the evolution of tighter levels of securitisation in specific places as a requirement of secrecy for military-industrial development would be another. A further example would be the move towards outsourcing and sub-contracting of formerly military functions to the private sector. Encouraged through the adoption of neo-liberal regimes of economic management, such changes have profound potential consequences for our understanding of what constitutes 'military' and thus a 'military landscape'. The UK practice of outsourcing the management of army training areas to a private sector consortium (Landmarc) or the exponential growth in the use of private security contractors who require basing and training facilities quite distinct from those habitually developed by military forces both reflect changes on landscapes which, through their very presence, blur a more traditional distinction between the military and civilian spheres.

Finally, we should consider the agendas suggested by the changing nature of civil-military relations, brought into being by changing structures and functions of armed forces and reflected in shifts in the ways in which civil society responds to its armed forces. New modes of commemoration and mourning emergent during the Iraq and Afghanistan wars and prompted by the public repatriation of the bodies of fallen soldiers illustrate this well. As Walklate, Mythen, and McGarry (2011) note, the solemn silent crowds lining the route taken by hearses carrying the bodies of fallen soldiers through the Wiltshire town of Wootton Bassett denotes the emergence of a reconfigured politics of remembrance which is simultaneously appropriate and uncomfortable because of the questions these acts pose for our acceptance (or otherwise) of the legitimacy of conflict. New modes of civilian regulation of activities which impact upon the environment, which require adherence following the outsourcing of military functions to civilian management, have yet to be investigated, but these too raise difficult questions about public acceptance (or otherwise) of the costs of military activities. Clearly, then, there is scope not just for continued sustained examination of the nature of military activities and institutions and their relationship to the civil societies in which they are based, but also for critique of the consequences of such activities as they are written across the landscape.

REFERENCES

Bushaway, B. (1992). Name upon name: The Great War and remembrance. In R. Porter (Ed.), *Myths of the English* (pp. 136–167). Cambridge: Polity Press.

Cosgrove, D., & Daniels, S. (Eds.). (1988). *The iconography of landscape: Essays on the symbolic representation, design and use of past environments*. Cambridge: Cambridge University Press.

Cowan, D., & Smith, N. (2009). After geopolitics? From the geopolitical social to geoeconomics. *Antipode, 41*(1), 22–48.

Davis, J. S. (2005). Representing place: 'Deserted isles' and the reproduction of Bikini Atoll. *Annals of the Association of American Geographers, 95*, 607–625.

Davis, J. S. (2007). Scales of Eden: Conservation and pristine devastation on Bikini Atoll. *Environment and Planning D: Society and Space, 25*(2), 213–235.

Davis, J. S., Hayes-Conroy, J. S., & Jones, V. M. (2007). Military pollution and natural purity: Seeing nature and knowing contamination in Vieques, Puerto Rico. *GeoJournal, 69*, 165–179.

Dimitrova, S. (2005). 'Taming the death': The culture of death (1915–1918) and its remembering and commemorating through First World War soldier monuments in Bulgaria (1917–1944). *Social History, 30*(2), 175–194.

Doyle, P., & Bennett, M. (Eds.). (2002). *Fields of battle: Terrain in military history*. Amsterdam: Kluwer.

Ferguson, K. E., & Turnbull, P. (1998). *Oh, say, can you see? The semiotics of the military in Hawaii*. Minneapolis, MN: University of Minnesota Press.

Foster, J. (2004). Creating a *temenos*, positing 'South Africanism': Material memory, landscape practice and the circulation of identity at Delville Wood. *Cultural Geographies, 11*, 259–290.

Gough, P. (2004). Sites in the imagination: The Beaumont Hamel Newfoundland memorial on the Somme. *Cultural Geographies, 11*, 235–258.

Graham, S. (2009). Cities as battlespace: The new military urbanism. *City, 13*(4), 383–402.

Graham, S. (2010). *Cities under siege: The new military urbanism*. London: Verso.

Graham, S. (2012). When life itself is war: On the urbanization of military and security doctrine. *International Journal of Urban and Regional Research, 36*(1), 136–155.

Havlick, D. (2007). Logics of change for military-to-wildlife conversions in the United States. *GeoJournal, 69*, 151–164.

Havlick, D. (2011). Disarming nature: Converting military lands to wildlife refuges. *The Geographical Review, 101*(2), 183–200.

Heffernan, M. (1995). For ever England: The western front and the politics of remembrance in Britain. *Ecumene, 2*(3), 293–323.

Herman, R. D. K. (2008). Inscribing empire: Guam and the war in the Pacific National Historical Park. *Political Geography, 27*, 630–651.

Higate, P., & Henry, M. (2011). Militarising spaces: A geographical exploration of Cyprus. In S. Kirsch & C. Flint (Eds.), *Reconstructing conflict: Integrating war and post-war geographies* (pp. 133–155). Farnham: Ashgate.

Hucker, J. (2009). 'Battle and Burial': Recapturing the cultural meaning of Canada's national memorial on Vimy Ridge. *The Public Historian, 31*(1), 89–109.

Iles, J. (2003). Death, leisure and landscape: British tourism to the Western Front. In M. Dorrian & G. Rose (Eds.), *Deterritorialisations: Revisioning landscapes and politics* (pp. 234–243). London: Black Dog Publishing.

Lahiri, N. (2003). Commemorating and remembering 1857: The revolt in Delhi and its afterlife. *World Archaeology, 35*(1), 35–60.

MacDonald, F., Hughes, R., & Dodds, K. (Eds.). (2010). *Observant states: Geopolitics and visual culture*. London: IB Tauris.

Masson, E., & Harden, J. (2009). Drumossie Moor: Memorialisation, development and restoration in an evolving historic landscape. In T. Pollard (Ed.), Culloden: The history and archaeology of the last clan battle (pp. 203–217). Barnsley: Pen and Sword.

Morris, M. S. (1997). Gardens 'For Ever England': Landscape, identity and the First World War British cemeteries on the Western Front. *Ecumene, 4*(4), 410–434.

Muir, R. (1999). *Approaches to landscape*. Basingstoke: Macmillan.

Muzaini, H. (2006). Producing/consuming memoryscapes: The genesis/politics of Second World War commemoration in Singapore. *GeoJournal, 66*(3), 211–222.

Muzaini, H., & Yeoh, B. (2005a). War landscapes as 'battlefields' of collective memories: Reading the *reflections at Bukit Chandu*, Singapore. *Cultural Geographies, 12*, 345–365.

Muzaini, H., & Yeoh, B. (2005b). Contesting 'local' commemoration of the Second World War: The case of the Changi chapel and museum in Singapore. *Australian Geographer, 36*(1), 1–17.

Muzaini, H., & Yeoh, B. (2005c). Reading representations of women's war experiences in the Changi chapel and museum, Singapore. *Geoforum, 36*, 465–476.

Muzaini, H., & Yeoh, B. (2007). Memory-making 'from below': Rescaling remembrance at the Kranji war memorial and cemetery, Singapore. *Environment and Planning A, 39*(6), 1288–1305.

Paglen, T. (2009). *Blank spots on the map. The dark geography of the Pentagon's secret world.* New York, NY: Dutton.

Palka, E. J., Galgano, F. A., & Corson, M. W. (2005). Operation Iraqi freedom: A military geographical perspective. *Geographical Review, 95*(3), 373–399.

Passmore, D., & Harrison, S. (2008). Landscapes of the battle of the Bulge: WW2 field fortifications in the Ardennes forests of Belgium. *Journal of Conflict Archaeology, 4*(1–2), 87–108.

Pearson, C. (2006). 'The Age of Wood': Fuel and fighting in French forests, 1940–1944. *Environmental History, 11*, 775–803.

Pearson, C. (2007). L'âge du bois: Les forêts françaises pendant la Seconde guerre mondiale. *Revue forestière française, 4*, 393–416.

Pearson, C. (2008). *Scarred landscapes: War and nature in Vichy France*. Basingstoke: Palgrave Macmillan.

Pearson, C. (2009). A 'Watery Desert' in Vichy France: The environmental history of the Camargue Wetlands, 1940–1944. *French Historical Studies, 32*(3), 479–509.

Pearson, C. (2012). *Mobilizing nature: The environmental history of war and militarization in modern France*. Manchester: Manchester University Press.

Pearson, C., Coates, P., & Cole, T. (Eds.). (2010). *Militarized landscapes: From Gettysburg to Salisbury Plain*. London: Continuum.

Peluso, N. L., & Vandergeest, P. (2011). Political ecologies of war and forests: Counterinsurgencies and the making of national natures. *Annals of the Association of American Geographers*, *101*(3), 587–608.

Perkins, C., & Dodge, M. (2009). Satellite imagery and the spectacle of secret spaces. *Geoforum*, *40*, 546–560.

Pollard, T. (2007). Burying the hatchet? The post-combat appropriation of battlefield spaces. In L. Purbrick, J. Aulich, & G. Dawson (Eds.), *Contested spaces: Sites, representations and histories of conflict* (pp. 121–145). Basingstoke: Palgrave Macmillan.

Pollard, T. (2009). Capturing the moment: The archaeology of Culloden battlefield. In T. Pollard (Ed.), *Culloden: The history and archaeology of the last clan battle* (pp. 130–162). Barnsley: Pen and Sword.

Rainbird, P. (2003). Representing nation, dividing community: The Broken Hill war memorial, New South Wales, Australia. *World Archaeology*, *35*(1), 22–34.

Raivo, P. J. (2004). 'This is where they fought': Finnish war landscapes as a national heritage. In T. Ashplant, G. Dawson, & M. Roper (Eds.), *Commemorating war* (pp. 145–164). New Brunswick, NJ: Transaction.

Smith, R. (2006). *The utility of force: The art of war in the modern world*. Harmondsworth: Penguin.

Switzer, C., & Graham, B. (2010). 'Ulster's love in letter'd gold': The battle of the Somme and the Ulster Memorial Tower, 1918–1935. *Journal of Historical Geography*, *36*, 183–193.

Tivers, J. (1999). 'The Home of the British Army': The iconic construction of military defence landscapes. *Landscape Research*, *24*(3), 303–319.

Walklate, S., Mythen, G., & McGarry, R. (2011). Witnessing Wootton Bassett: An exploration in cultural victimology. *Crime Media Culture*, *7*(2), 149–165.

Weizman, E. (2007). *Hollow land: Israel's architecture of occupation*. London: Verso.

Westing, A. (1988). The military sector vis-à-vis the environment. *Journal of Peace Research*, *25*(3), 257–264.

Woodward, R. (2004). *Military geographies*. Oxford: Blackwell.

Woodward, R. (2005). From military geography to militarism's geographies: Disciplinary engagements with the geographies of militarism and military activities. *Progress in Human Geography*, *29*(6), 718–740.

Wylie, J. (2007). *Landscape*. London: Routledge.

Yamazaki, T. (2011). The US militarization of a 'host' civilian society: The case of postwar Okinawa, Japan. In S. Kirsch & C. Flint (Eds.), *Reconstructing conflict: Integrating war and post-war geographies* (pp. 253–272). Farnham: Ashgate.

PART III
BUILDING UP THE HUMAN RESOURCES OF ARMED FORCES

THE IMPACT OF MILITARY SERVICE ON FUTURE LABOR-MARKET OUTCOMES

Alfredo R. Paloyo

ABSTRACT

The purpose of this chapter is to illustrate the identification problem involved in estimating the impact of military service on labor-market outcomes and to review the modern literature in economics which aim to estimate this effect. Drawing from the literature on microeconometric treatment evaluation, the existence of a selection bias is demonstrated and a discussion of empirical strategies to overcome this is presented. Studies on the effect of military service in the United States, Germany, the Netherlands, Portugal, and the United Kingdom are discussed. The main results of each reviewed article are presented. There is substantial heterogeneity in the estimated impacts of military service on labor-market outcomes. The results may not be generalizable to the general population due to the econometric methods employed. Moreover, the literature review is concentrated on the contribution of economists to this issue and neglects to discuss research conducted by other social scientists. The studies are varied and located in different journals. This review serves to summarize the results in one article. Furthermore, the technical

The Evolving Boundaries of Defence: An Assessment of Recent Shifts in Defence Activities
Contributions to Conflict Management, Peace Economics and Development, Volume 23, 157–176

ISSN: 1572-8323/doi:10.1108/S1572-832320140000023011

discussion on methods will be useful for those who wish to pursue the topic further. This chapter provides practical advice on how to credibly estimate treatment effects based on nonexperimental data. It also facilitates future reviews of the topic by collecting the available evidence.

INTRODUCTION[1]

The nexus between military service and the subsequent civilian labor-market performance of former soldiers has attracted and continues to attract careful analysis from social scientists. The interface between the two facets of a veteran's life – military and, eventually, civilian – is quite complex and richly deserves the attention it has received thus far. Indeed, the abundance and variety of reported effects of military service reflect the intricate nuances inherent to this issue.

One way to resolve the apparent contradictions when comparing specific estimated impacts is to realize that the potential effect of military service on later success in the labor market turns on the prevailing institutional mechanisms that soldiers and conscripts face both during and after active duty. For example, Angrist (1990) reports that Vietnam War white veterans earned 15 percent less than comparable nonveterans. In contrast, Angrist and Krueger (1994) fail to find similar effects for World War II veterans. Certainly, the conditions faced by Vietnam veterans were different from those experienced by their World War II counterparts both abroad and at home.

At the outset, it is important to distinguish between wartime and peacetime military service. The vast majority of the empirical studies on military service have concentrated on the effect of serving in World War II or the Vietnam War on various economic and demographic outcomes.[2] Recently, however, peacetime compulsory military service has drawn considerable attention.[3] The expected impact of these two types of military service is not necessarily the same for a variety of reasons, not the least of which is the fact that combat (wartime) duty necessarily increases the risk of death.

Moreover, a distinction has to be made between compulsory military service and volunteers to the armed forces. The characteristics of entering cohorts of servicemen and -women are different under the two staffing regimes of the armed forces. People may enlist voluntarily in the armed forces when the prospects for civilian labor-market success are dim. Under conscription,[4] even those who may be better off in the civilian labor market may not be able to avoid serving in the military. In other words, volunteers self-select themselves into the armed forces while conscripts generally do not.

Although the ways in which conscription is implemented are as varied as the number of countries that practice it, the general procedure is that men[5] are required to undergo physical and mental examinations once they reach the age of majority to determine whether they are fit to serve in the armed forces. Those who are not healthy enough or are mentally incapacitated to adequately serve are exempted from service while the rest becomes eligible for the draft. Whether the men are actually called for service depends on the needs of the armed forces. Thus, the percentage of drafted men for a given cohort may vary over time.

We review a number of significant economic contributions to this literature spanning over half a century. We structure the rest of the chapter as follows. First, we enumerate the reasons that underpin the purported link between military service and civilian labor-market performance. In section "Identifying the Effect of Military Service," we discuss why estimating the effect of military service on labor-market outcomes is complicated by the endogeneity of military service itself. In this section, we also expound on the econometric techniques used to overcome the endogeneity problem. In section "Survey of the Modern Literature," we survey the existing literature and report the various estimates from the relevant studies. Finally, we conclude with a reflection on the potential direction of future research on this topic.

HOW CAN MILITARY SERVICE AFFECT LABOR-MARKET PERFORMANCE?

Whether military service can improve or diminish one's performance in the civilian labor market is not entirely clear a priori. For instance, while certain skills acquired during service may be valued later on upon exit from the armed forces (e.g., the ability to properly follow orders), soldiers on active duty are also unable to invest in firm-specific human capital. For this and a number of other reasons we elaborate on below, a generalized conclusion as to the effect of military service is haphazard, to say the least. As will be made clear in subsequent sections, the ultimate effect of military service can only be meaningfully interpreted within the context of a specific institutional and historical setting.

Benefits of Military Service

As previously mentioned, a soldier may acquire skills while on active duty that may be subsequently rewarded in the civilian labor market. In Germany, for instance, most conscripts benefited from the free driving lessons offered by the German armed forces (*Bundeswehr*) to acquire a driver's license. A more extreme case is faced by the Philippine Air Force, which lost 54 pilots to local and foreign commercial airlines in 2006.[6] Soldiers are also taught to operate in a hierarchical environment with a very specific chain of command. Although working for most firms certainly does not involve the hanging threat of doing pushups for every mishap, companies do have a hierarchical structure involving a fairly strict delineation of duties. Such a configuration should be familiar to former soldiers. Moreover, if a soldier subsequently chooses to work for a defense contractor or supplier, she can be expected to bring with her essential (perhaps even privileged) information that may be highly desired by her new employers (e.g., the strengths and weaknesses of the design of a competitor's products). Since working for the armed forces typically means being attached to a unit, a soldier may also learn to manage interpersonal relationships with teammates, a skill that is highly valued in a company.

Those with the least propensity to succeed in the formal educational system may prosper in an alternative setting offered by the armed forces. In the case of conscripts, for instance, they may even be forced to acquire marketable skills which they would not have otherwise developed on their own. For example, discipline is often cited as lacking in problematic youths and this is presumably inculcated by members of the armed forces.

Veterans may also be rewarded after their service and this may give them a distinct advantage over nonveterans. For instance, the G.I. Bill (Servicemen's Readjustment Act of 1944) offered incentives for further education and unemployment benefits for veterans of World War II. Bound and Tuner (2002) show that the G.I. Bill substantially increased the college attainment of World War II veterans. This increase in educational attainment may naturally translate to better performance in the labor market, considering the nontrivial returns to education (see, e.g., Harmon, Oosterbeek, and Walker, 2003).

Finally, the opportunities to network within the armed forces and the signal to employers for performing military service are difficult to underestimate. Veterans may be afforded preferential treatment in the labor market for having served their country. Peacetime service, for instance, may also be taken as a signal for superior health.[7] Capitalizing on network effects (or, perhaps more perniciously, nepotism), soldiers entering or new in the civilian labor market may also benefit from earlier veterans when the latter can influence the former's chances of employment, promotion, or receiving a pay raise, such as when the veteran is the superior of a newly hired former soldier employed at the same firm.

Costs of Military Service

In the previous section, we outlined the numerous benefits of military service. Despite these potential advantages, we fail to observe a massive rush of the population to enlist in the armed forces. The reason is that the cost of military service is substantial and real. Indeed, in the long run, it may far outweigh the benefits.

The most obvious cost of military service – specifically, compulsory military service – is the hidden tax that it levies on soldiers (Oi, 1967). This tax comes in the form of the opportunity cost of not immediately engaging the civilian labor market but rather being drafted into the armed forces instead. The wage (if any) in the military is generally less than what a conscript can receive in the civilian labor market. The difference in the military and the civilian wage is an implicit tax borne by the soldier.[8]

While the opportunity cost of military service is an important consideration, it is not the subject of this survey.[9] Instead, we select the studies that estimate the specific labor-market penalties that military service itself may generate. These penalties materialize because of the loss of labor-market experience, the depreciation of human capital, the effect of draft uncertainty on human-capital investments and employment probability, the adjustment cost associated with re-assimilation into civilian life, and the risk to the soldier's health.

The loss of civilian labor-market experience is crucial because associated with this loss is the forfeiture of the contribution of labor-market experience to wage growth, as evidenced by a substantial literature on the effects of labor-market experience and tenure on wage dynamics (see, e.g., Altonji & Williams, 1998). At this point, the soldier is unable to invest in firm-specific and general human capital and the substitute that she may acquire in the armed forces may be inappropriate for civilian purposes.

Entrants into the armed forces typically have finished at least secondary schooling. The entire stock of human capital that they bring into service may not be useful for the armed forces. Thus, part of the stock may depreciate because of underutilization (Spencer & Woroniak, 1969). The overall returns (in the form of wages) to human capital will be less than what the person could have earned if they had started working after high school or if they had pursued higher education.

Conscripts are unable to accurately predict when exactly they will be called for service after they have been declared fit to serve. Due to this uncertainty, they may find difficulties finding employment since the employer cannot be assured that her occupation-specific investment can be recouped since the investment recipient (the potential conscript) may be drafted at any time. Less permanent employment (and also, less-paying) may be more likely to be accepted by this group.

Grenet, Hart, and Roberts (2011) note that former conscripts face adjustment costs as they transition into civilian employment. They cite a report which states that companies encountered difficulties with the work motivation of former conscripts since the demand on an individual's working time is higher in these private companies.

Finally, the risk of injury or the risk to life is an obvious concern. Particularly for those who engage in combat duty, there is a nontrivial chance of getting injured. This occupational hazard is inherent to the performance of one's duty in the armed forces. Based on the United States, the 12-month prevalence rate for post-traumatic stress disorder (PTSD) is about 3.5 percent of the adult population. About 11–20 percent of the veterans of the Iraq War (Operation Iraqi Freedom) and the Afghanistan War (Operation Enduring Freedom) suffer from PTSD; 10 percent of the Gulf War (Desert Storm) veterans and 30 percent of Vietnam War veterans also suffer from PTSD.[10] Citing another city, McCrone, Knapp, and Cawkill (2003) mention that 22 percent of a sample of 64 Falkland War veterans have had PTSD. These and other mental or physical incapacitation reduces a person's ability to work and will likely have a negative impact on one's potential to succeed in the labor market.

IDENTIFYING THE EFFECT OF MILITARY SERVICE

Comparing the labor-market outcomes of those who served in the armed forces to those who did not serve does not necessarily lead to the effect of military service on such outcomes. For instance, Angrist and Krueger (1994) observe that World War II veterans earn more than nonveterans in the same cohort. The question arises: to what extent is this difference in earnings the exclusive result of military service? The answer is complicated by the fact that selection into military service is typically nonrandom. For a person to be considered fit to serve, he normally has to pass a physical fitness and a mental aptitude test. Simple comparisons between veterans and nonveterans fail to account for differences in these dimensions between both pools of men because this information is often unavailable to the researcher.

Why would accounting for physical fitness and mental aptitude matter? Health (physical and mental) is positively associated with labor-market outcomes (Smith, 1999; Strauss & Thomas, 1998). In other words, healthier and smarter people, on average, tend to perform better in terms of earnings and employment. Since selection into military service hinges on one's performance in these pretests, the pool of men who serve are characteristically different from those who do not serve, particularly in traits that simultaneously affect labor-market performance and the likelihood of military service. Without accurate information to correct for these differences, a straightforward comparison of mean outcomes will be uninformative as to the true effect of military service on subsequent labor-market outcomes.

Selection Bias

We formalize this selection problem as follows.[11] Suppose we have a sample of N men denoted by i. Our measure for labor-market performance, say, wages, is y_i. A binary indicator, m_i, equals 1 if the person served in the armed forces (i.e., was "treated") and 0 otherwise. Thus, before the assignment into the armed forces, there are two potential outcomes for each person: $y_i(0)$ and $y_i(1)$ which represent the outcomes if i did not serve or if i served, respectively.

What we are interested in is the difference between the two potential outcomes, which gives us the effect of military service for that specific person. On average, this is $E[y_i(1) - y_i(0)]$, where E is the mathematical expectation operator. However, we only observe the following from the data: $y_i = y_i(m_i) = y_i(0)\ (1 - m_i) + y_i(1)(m_i)$. That is, we can only observe one of the potential outcomes at a time because they are mutually exclusive. This is referred to as the "fundamental problem of causal inference" by Holland (1986).

From the available data, we can compute the difference in wages between veterans and nonveterans and relate it to the average effect of military service on those who served (average treatment effect on the treated or ATET) and the bias introduced by nonrandom selection:

$$\underbrace{E[y_i|m_i=1]-E[y_i|m_i=0]}_{\text{Observed difference}} = \underbrace{E[y_i(1)|m_i=1]-E[y_i(0)|m_i=1]}_{\text{ATET}} + \underbrace{E[y_i(0)|m_i=1]-E[y_i(0)|m_i=0]}_{\text{Selection bias}} \quad (1)$$

Note that the observed difference is equal to the ATET if and only if there is no selection bias in Eq. (1). With randomized treatment assignment, such as what is found typically in experimental settings, the selection bias disappears by design.

With respect to military service, the selection-bias term is generally nonzero because the probability of service and labor-market performance are both affected by unobserved variables such as health and mental aptitude. Under a regime of compulsory military service, those who are drafted into the armed forces tend to be healthier and more intelligent than the nondraftees. Since we can expect the healthier and smarter to perform better in the labor market, it is likely that $E[y_i(0)|m_i=1]>E[y_i(0)|m_i=0]$. Therefore, simply using the observed difference in y_i between veterans and nonveterans will overestimate the impact of military service on wages.

The previous paragraph describes a situation where positive selection prevails. However, an alternative institutional setting may induce negative selection instead. An all-volunteer force would also make comparisons between veterans and nonveterans somewhat more involved because of selection biases but the reason may not be that volunteers to the armed forces are smarter and healthier than those who refused to volunteer. The resulting entrants under such a self-selection mechanism are rather quite the opposite of what we can expect from those who were mandatorily drafted. For example, men and women who are more likely to fare poorly in the civilian labor market may opt to join the armed forces to secure employment. In this case, a simple comparison of mean outcomes in wages may overestimate the negative impact of military service.

Empirical Strategies

How have economists overcome the problem associated with selection bias in the absence of randomized controlled trials? In the literature, a number of researchers have relied on the instrumental-variables (IV) approach to credibly identify the effects of military service on various socioeconomic outcomes.[12] When appropriate, others have exploited the regression-discontinuity (RD) design of compulsory military service, such as for the United Kingdom and for Germany. Both the IV and RD approaches identify a specific type of average treatment effect that is important to note when discussing the generalizability of the results presented in such studies.

Instrumental Variables

One can apply the instrumental-variable approach if there exists a variable, say, z_i, that is exogenous to the potential outcome (i.e., does not affect the potential outcome independently) but affects the probability of being treated. We call z_i an instrumental variable. We now have a quintuple $\{z_i, m_i(0), m_i, (1), y_i(1), y_i(0)\}$, where m_i is a function of z_i and, as before, y_i is a function of m_i. Assuming that the instrument is binary, the observed treatment is $m_i = m_i(0)(1 - z_i) + m_i(1)(z_i)$. The validity of the instrument rests on the assumption that it does not have a direct effect on the potential outcome: $\{[y_i(0), y_i(1), m_i(0), m_i(1)] \perp z_i\}$, where $\perp$ represents statistical independence. Along with the assumption of monotonicity, the IV approach recovers the local average treatment effect (LATE).[13] The identified treatment effect is called "local" because it is valid only for a particular subpopulation called the "compliers."

In principle, there are four groups of people based on compliance type: the never-takers, the compliers, the defiers, and the always-takers. The never-takers are those who will not receive treatment irrespective of what z_i is; in contrast, the always-takers are those who will receive treatment irrespective of what z_i is. The compliers are those who will not receive treatment if z_i is zero but will receive treatment if z_i is one; the defiers are those who will receive treatment if z_i is zero but will not receive treatment if z_i is one. Denote c_i as the compliance type. Then, the above can be written compactly as:

$$c_i = \begin{cases} \text{Never-taker if } m_i(0) = 0, \quad m_i(1) = 0; \\ \text{Complier if } m_i(0) = 0, \quad m_i(1) = 1; \\ \text{Defier if } m_i(0) = 1, \quad m_i(1) = 0; \\ \text{Always-taker if } m_i(0) = 1, \; m_i(1) = 1. \end{cases}$$

By the assumption of monotonicity, the existence of defiers is ruled out.

Under these conditions, the IV approach estimates a specific treatment effect, the LATE: $E[y_i(1)-y_i(0)|m_i(0)=0, m_i(1)=1]$ or the average treatment effect for the compliers in the population. This can be estimated via two-stage least squares (2SLS) or by using Wald's (1940) grouping estimator.

When estimating the LATE of military service using 2SLS, the first- and second-stage appear as follows:

$$m_i = \alpha + \beta z_i + \boldsymbol{\xi}'\mathbf{X}_i + \epsilon_i \tag{2.1}$$

$$y_i = \gamma + \tau \hat{m}_i + \boldsymbol{\chi}'\mathbf{X}_i + \nu_i, \tag{2.2}$$

where τ represents the effect of m_i (say, military service) on wages, y_i, for the subpopulation of compliers; $\hat{m}_i$ is the predicted value based on Eq. (2.1), α, β, $\boldsymbol{\xi}$, γ, and $\boldsymbol{\chi}$ are (vectors of) parameters to be estimated, $\mathbf{X}_i$ is a vector of control variables, and ϵ_i and v_i are stochastic disturbance terms. The content of z_i will vary depending on the study design but it is crucial that it satisfies the exclusion restriction (i.e., that it is a "valid instrument": a variable that does not directly affect the outcome y_i) and that it is (strongly) predictive for the endogenous variable m_i.

The 2SLS approach is numerically equivalent to the grouping estimator:

$$\hat{\tau} = \frac{\left((1/N_{z_i=1})\sum_{i|z_i=1}^{N_{z_i=1}} y_i\right) - \left((1/N_{z_i=0})\sum_{i|z_i=0}^{N_{z_i=0}} y_i\right)}{\left((1/N_{z_i=1})\sum_{i|z_i=1}^{N_{z_i=1}} m_i\right) - \left((1/N_{z_i=0})\sum_{i|z_i=0}^{N_{z_i=0}} m_i\right)}. \tag{3}$$

The first term in the numerator is the mean wages computed from those people for which z_i is equal to one. We subtract the corresponding mean based on the subsample for which z_i is equal to zero. This difference in outcomes is then rescaled by dividing it by the compliance rate.

RD Design

The RD design requires the existence of a continuous "forcing variable" which acts like an instrument (Angrist & Pischke, 2009). The discontinuity arises because, once the forcing variable crosses a known threshold, the conditional probability of being treated jumps correspondingly. For instance, in Bauer, Bender, Paloyo, and Schmidt (2012) and Grenet et al.

(2011) discussed below, a function of the date of birth is used as an "instrument" for being drafted into the armed forces. The two studies exploit the institution in Germany and the abolition in the United Kingdom, respectively, of compulsory military service.

Denote the forcing variable as b_i and the threshold as $\bar{b}$. Then, similar to Eq. (3), the treatment effect can be estimated using the following (Hahn, Todd, & van der Klaauw, 2001):

$$\hat{\tau} = \frac{\lim_{b \downarrow \bar{b}} E[y_i | b_i = b] - \lim_{b \uparrow \bar{b}} E[y_i | b_i = b]}{\lim_{b \downarrow \bar{b}} E[m_i | b_i = b] - \lim_{b \uparrow \bar{b}} E[m_i | b_i = b]}. \tag{4}$$

The limits in Eq. (4) can be estimated in various ways (e.g., via kernel estimators). However, since b_i is essentially an instrument, one can also estimate the effect by 2SLS. The crucial difference between Eqs. (3) and (4) is that the identification of the treatment effect in the latter is based on observations that are close to the threshold point $\bar{b}$. In this sense, therefore, the RD estimand is particularly valid only for those observations that are rather close to the point of discontinuity.

SURVEY OF THE MODERN LITERATURE

It is convenient to divide the literature into papers that examine peacetime conscription and those that involve veterans of the Vietnam War and World War II. The occupational environment faced by veterans of wars was much more dangerous than those who had to do their national service as part of a country's institution of compulsory military service.

Service in Vietnam and World War II

Angrist (1990) notes that Vietnam veterans were earning less than nonveterans in the 1970s and early 1980s. On the other hand, World War II veterans seem to be earning more than nonveterans (Angrist & Krueger, 1994). Recognizing the selection bias that could occur in both situations (and hence prevent drawing definitive conclusions from these simple observations), the authors implemented an instrumental-variable approach to isolate the effect of military service in these wars on future earnings.

Analyzing the effect of military service on labor-market outcomes for Vietnam-era veterans is facilitated by the draft lottery that was instituted by the United States to post men into the armed forces for this war. The lottery mechanism went as follows. For the years 1969, 1970, and 1971, the lottery held in that year applied to the cohort of men born during the period/year 1944–1950, 1951, and 1952, respectively. Lindo and Stoecker (2014) describe it further:

> In each drawing, the birthdays of the year were randomly assigned a Random Sequence Number (RSN). In the 1969 drawing September 1st was assigned RSN 1 so men born on September 1st were asked to report to their local draft boards for potential induction before men born on other days. April 24th was assigned RSN 2 so men born on that day were asked to report second, and so forth. The military continued to call men for potential induction in order of RSN until the manpower requirements were met for that year. The last RSN called for service, also known as the highest Administrative Processing Number (APN), was 195 for the 1969 drawing, 125 for the 1970 drawing, and 95 for the 1971 drawing.

As for the "selective service" in World War II, Angrist and Krueger (1994) explain it as follows:

> There were a total of seven national draft registrations between 1940 and 1947. Virtually all male U.S. citizens and resident aliens were required to register … In the fifth and sixth registrations, held from June 1942 to March 1947, order numbers were assigned on the basis of birthdays, with individuals born in the beginning of the year more likely to be called for induction. The sixth registration was divided into two parts, with the second part covering the period 1943–47. The second part of the sixth registration, which included men born between January 1, 1925, and March 31, 1929, is significant because voluntary enlistments were prohibited during this registration.

In both cases, what is essentially generated by the draft mechanism is an instrumental variable that determines the probability of military service but is excludable from the regression of earnings on military service. For instance, denote having an RSN below the APN as $d_i = 1$ and zero otherwise for the Vietnam-era draft. Then, if we specify earnings as a linear function of having served in the armed forces ($m_i = 1$), our main equation of interest will look like $y_i = \alpha + \tau\hat{m}_i + \epsilon_i$, where $\hat{m}_i$ represents the predicted value from a first-stage regression of the following form: $m_i = \gamma + \beta d_i + v_i$. This allows us to consistently estimate the effect of military service on earnings since, by design (randomized lottery assignment), d_i is orthogonal to ϵ_i.

Using the above approach to correct for selection bias, Angrist (1990) reports that white veterans of the Vietnam War incur an earnings penalty of about 15 percent of annual wage and salary earnings based on their earlier census data. The explanation he offers is that military service is only a partial substitute to civilian labor-market experience. However, more recent estimates by Angrist, Chen, and Song (2011) indicate that this penalty has largely dissipated in later periods. Nevertheless, this fact is consistent with the loss-of-experience interpretation earlier offered and the estimates in Angrist and Chen (2011) based on more recent census data.

A similar approach is used to scrutinize the earnings premium of World War II veterans based on social security data. Angrist and Krueger (1994) used a vector of quarter-of-birth indicators and quarter-of-birth indicators interacted with year-of-birth indicators as an excluded instrument set for serving in the armed forces. The authors find that World War II veterans in fact do not earn more than comparable nonveterans. They attribute the observed wage premium to the fact that screening for military service tended to sort out the relatively unfit individuals from active duty.

Peacetime Conscription

While the studies in the previous section involve men who are veterans of wars, this section concentrates on the effects of peacetime conscription on subsequent labor-market outcomes. This issue has been analyzed in Germany, the Netherlands, Portugal, and the United Kingdom.

Germany

When West Germany joined NATO, the parliament passed a law to introduce compulsory military service to staff its newly reconstituted armed forces. This law stated that all men born on or after July 1, 1937, had to serve in the *Bundeswehr* while those who were born earlier were totally exempted from conscription. This specific study design is an example of a regression discontinuity design. Note that only men born on or after a particular threshold point (the date of birth July 1, 1937) faced a positive probability of being inducted while those born earlier were exempted. This is illustrated clearly in Fig. 1, which shows the share of conscripts by date of birth.

Using data from matched social security administration and pension authority records, Bauer et al. (2012) report that conscripts have a lifetime earnings[14] advantage of about 17 percent over nonconscripts. However, when exploiting the RD design of conscription to eliminate selection biases, the authors report that there is no statistically significant difference in earnings between conscripts and comparable nonconscripts.[15]

Essentially, the groups of men being compared are those men born just before and just after the threshold birth date. The assumption here is that

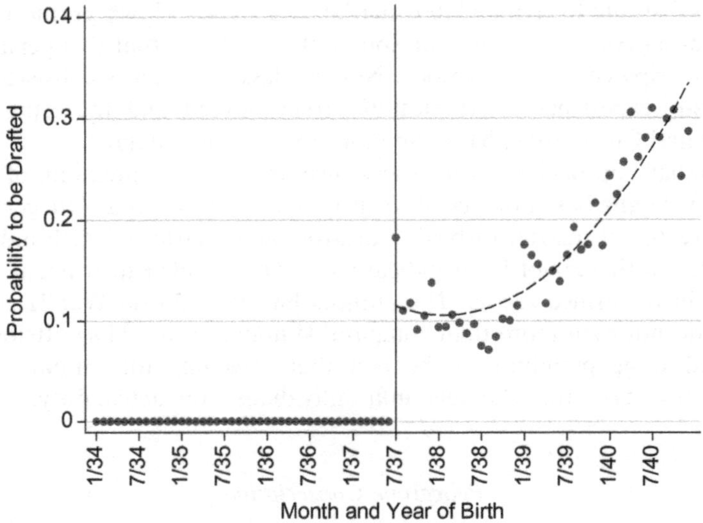

Fig. 1. Probability to Be Drafted for Military Service. *Notes*: The dashed lines represent the quadratic fit over the observation points. These are separately estimated for both sides of the threshold point. *Source*: Bauer et al. (2012).

the only thing that distinguishes these two groups from each other is the fact that those born later could have been drafted into the *Bundeswehr*. Similar to Angrist and Krueger (1994), they appeal to the fact that the screening mechanism for the *Bundeswehr* involved exemptions for those who were declared to be physically or mentally unfit to serve. This naturally meant that the more able men, who are more likely to succeed in the civilian labor market, anyway, were precisely the ones who were drafted into the *Bundeswehr*.

The Netherlands

Imbens and van der Klaauw (1995) are able to exploit the variability in conscription rates over time as an instrument for being drafted into the Dutch armed forces (*Krijgsmacht*) using data from the Dutch Defense Department. Earnings in 1989 and 1990 are obtained from data provided by the Central Bureau for Statistics in the Netherlands.

The idea behind this approach is that, while the individual's chances of being inducted are plagued with selection effects, the conscription rate for an entire cohort is not. Given a high enough level of aggregation, the selection bias is eliminated.

The authors conclude that, ten years after service, the annual cost of conscription is about 5 percent of earnings. This is comparable to losing about one year of civilian labor-market experience.

Portugal

Using administrative data called *Quadros de Pessoal*, Card and Cardoso (2012) provide evidence that the impact of conscription is likely to be heterogeneous. They show that the less-educated men benefit from performing military service but that such a service does not have any significant impact on men who are more educated. This study is also unique in that they do not use any of the empirical strategies mentioned in section "Empirical Strategies". Instead, they are able to control for a person's productivity before conscription using pre-enlistment wages because Portuguese conscripts were inducted relatively later in life.

During the 1980s, men in Portugal were required to serve in the military for a period of up to two years. This commenced when they reached 21 years of age, after having gone through a medical and psychological exam at 20. Like other screening mechanisms, those who were physically and mentally unfit were exempted from service. This late conscription age (compared to other countries, which typically induct their conscripts around the year they turn 18) means that some men have already started working. Assuming that wages reflect a worker's productivity, pre-service wages can be used to control for innate differences among men.

The authors report that they find insignificant effects of conscription on wages on average. However, they further demonstrate that men with lower levels of education (defined as those with less than six years of schooling) benefit from conscription by about 4–5 percent. Men with higher education do not show any statistically significant effect.

In this case, it seems that those men who were not likely to perform well in the civilian labor market because of their poor investments in human capital benefited the most from military service. They may have acquired certain skills in the armed forces (a possibility described in section "Benefits of Military Service") that they readily applied when they exited the armed forces. For these people, it seems that the training offered by the military is able to substitute for their lack of formal education.

The United Kingdom

Studies for the United Kingdom are similar to Bauer et al. (2012) for Germany because they also exploit the RD design of compulsory military service. The difference is that Bauer et al. (2012) use the introduction of conscription in Germany while Buonanno (2006) and Grenet et al. (2011) use its abolition in the United Kingdom. National service was abolished in the United Kingdom in 1960 and men born in 1939 comprised the last cohort which faced conscription. More precisely, those born after September 1939 were no longer required to report for national service (Grenet et al., 2011).

The two studies on the United Kingdom find conflicting results, with Buonanno (2006) reporting that conscripts earned about 5–8 percent less than nonconscripts while Grenet et al. (2011) show that there are no statistically significant long-term real earnings differences between the two groups of men. The earlier result is consistent with the estimates produced by Imbens and van der Klaauw (1995) for the Netherlands while the latter is more in line with the results for Germany.

The contradictory findings may be the result of the use of different data sources. Buonanno (2006) uses the UK General Household Survey (GHS) for the period 1983–1998 and the UK Labour Force Survey for the period 1984–1991. Grenet et al. (2011) use the New Earnings Survey Panel Data from 1975 to 2001, the GHS for the period 1979–1993, and the UK Family Expenditure Survey from 1982 to 1993.

CONCLUSION AND OUTLOOK

Military service is still an important aspect of many people's lives. A significant number of countries in the world still implement some form of compulsory military service. Those countries which have transitioned to an all-volunteer force still face many defense and strategic challenges that require the services of its armed forces. With the globalized nature of national defense today, the relevance of examining the unintended consequences of military service on measures of well-being (broadly defined to include, among others, labor-market performance, health status, and educational attainment) becomes even stronger.

Ideally, a researcher can estimate the effect of military service by simply randomizing military service, ensuring perfect compliance, and comparing subsequent outcomes. This is not what has happened thus far. Moreover, isolating the impact of military service on such outcomes is complicated by the lack of information before and after service.

In response to such a deficiency, economists have used econometric techniques to credibly recover the effect of military service despite the absence of experimental data. There are some advantages to this approach, not the least of which is that it dispenses with the need for running an experiment, which, if not patently unethical, is, in most cases, prohibitively expensive. There are, however, obvious disadvantages: the lack of generalizability being a prominent one.[16]

This lack of generalizability of an IV or an RD estimate to the general population is reflected in the variety of results reported above. A substantial part of this can be explained by the fact that institutional and historical settings are different in most of the studies above. For example, combat duty cannot be reasonably expected to be similar to peacetime conscription. Despite this, an obvious heterogeneity persists in the estimated effects of military service. To illustrate, Card and Cardoso (2012) find a positive effect for less-educated men but do not find a similar effect for men with higher education using data from Portugal. In contrast, Imbens and van der Klaauw (1995) report an annual cost of conscription of about 5 percent, equivalent to a loss of one year's civilian labor-market experience.

Do these apparently conflicting pieces of evidence doom this research agenda? On the contrary, it calls for even more specific studies that explicitly distinguish the institutional setting in which it finds itself. Note that even older sets of estimates are being revisited (e.g., Angrist, 1990; Angrist et al., 2011; Bauer et al., 2012; Paloyo, 2010). The importance of replication – here, we mean not merely the repetition of a study using the same methods and the same dataset (although that is also important for scientific integrity) but rather the repetition of the same exercise based on perhaps a different dataset or another institutional setting – is difficult to overstate. Estimating the impacts of military service on different subpopulations, like the approach undertaken in Card and Cardoso (2012) and in the various robustness checks in other studies, may reveal hidden insights that are masked by simply looking at the average.

That, at least, is an invitation extended to other researchers who wish to contribute to and shape the nation's defense policy. One possible avenue of future research is the effect on wages and employment of the sudden increase in the supply of male labor after wartime conscription was abolished.

NOTES

1. This chapter was written while the author was still with the Rheinisch-Westfälisches Institut für Wirtschaftsforschung (RWI) in Essen, Germany.

2. On this, the reader is referred to the long line of research (two of which have already been cited) produced by Joshua D. Angrist and his coauthors. Some of these are discussed in detail in section "Service in Vietnam and World War II."

3. The countries that have been examined thus far in this respect are Germany (Bauer, Bender, Paloyo, & Schmidt, 2011, 2012; Paloyo, 2010), the United Kingdom (Buonanno, 2006; Grenet, Hart, & Roberts, 2011), the Netherlands (Imbens & van der Klaauw, 1995), and Portugal (Card & Cardoso, 2012). Keller, Poutvaara, and Wagener (2009), Keller, Poutvaara, and Wagener (2010), Lau, Poutvaara, and Wagener (2004), and Poutvaara and Wagener (2007) develop models that explain the relationship between the military draft and economic growth and the demand for higher education. They also test the predictions of their models against data from OECD countries.

4. Throughout this chapter, compulsory military service and conscription are used interchangeably.

5. Women are not typically drafted although there are a number of notable exceptions, including (but not limited to) Israel, China, and North Korea.

6. Malig (2011). The problem is so severe that the Philippine Air Force is considering extending the eight-year service contract by another two years to protect their investments in training. See "PAF mulls longer service contracts for its pilots" by Victor Reyes (2011).

7. This may not apply for wartime service because of the prevalence of post-traumatic stress disorder in the armed forces.

8. This is unlikely the case for volunteers because, from their point of view, the expected returns – monetary and nonmonetary – in the civilian labor market are less than what they can receive in the armed forces.

9. However, as an example, note that Schleicher (1996) estimates that the average monthly opportunity cost for a German conscript is €1,950.

10. These figures come from the US Department of Veteran Affairs (2011).

11. This formalization draws from Imbens and Wooldridge (2009) and Angrist and Pischke (2009).

12. Apart from typical labor-market outcomes, health and education have also been examined.

13. In this case, the monotonicity assumption takes the form $m_i(1) \geq m_i(0)\ \forall\ i$ or, in words, each unit responds similarly to the instrument. For more details with respect to the identification assumptions behind this approach to recovering the treatment effect, the reader is referred to Imbens and Angrist (1994).

14. Bauer et al. (2012) also examine the impact of compulsory military service on lifetime average daily wages and the cumulative days of employment. We report only the results for earnings here and refer the interested reader to the original source.

15. This finding is reinforced by Paloyo (2010) using an alternative identification strategy of comparing men and women born before and after the threshold date.

16. In the literature, this is referred to as the "external validity" of the estimated effect.

ACKNOWLEDGMENT

The author wishes to thank Renaud Bellais for constructive comments and suggestions and Marion Dugué for editorial assistance.

REFERENCES

Altonji, J. G., & Williams, N. (1998). The effects of labor market experience, job seniority, and job mobility on wage growth. *Research in Labor Economics*, *17*, 233–276.

Angrist, J., & Krueger, A. B. (1994). Why do World War II veterans earn more than nonveterans? *Journal of Labor Economics*, *12*(1), 74–97.

Angrist, J. D. (1990). Lifetime earnings and the Vietnam era draft lottery: Evidence from social security administrative records. *American Economic Review*, *80*(3), 313–336.

Angrist, J. D., & Chen, S. H. (2011). Schooling and the Vietnam-era GI Bill: Evidence from the draft lottery. *American Economic Journal: Applied Economics*, *3*(2), 96–119.

Angrist, J. D., Chen, S. H., & Song, J. (2011). Long-term consequences of Vietnam-era conscription: New estimates using social security data. *American Economic Review*, *101*(3), 334–338.

Angrist, J. D., & Pischke, J.-S. (2009). Mostly harmless econometrics: An empiricist's compani*on*. Princeton, NJ: Princeton University Press.

Bauer, T. K., Bender, S., Paloyo, A. R., & Schmidt, C. M. (2011). Do guns displace books? The impact of compulsory military service on educational attainment, Ruhr Economic Paper No. 260. Retrieved from http://www.rwi-essen.de/publikationen/ruhr-economic-papers/379/

Bauer, T. K., Bender, S., Paloyo, A. R., & Schmidt, C. M. (2012). Evaluating the labor-market effects of compulsory military service. *European Economic Review*, *56*(4), 814–829.

Bound, J., & Turner, S. E. (2002). Going to war and going to college: Did the World War II and the G.I. Bill increase educational attainment for returning veterans? *Journal of Labor Economics*, *20*(4), 784–815.

Buonanno, P. (2006). *Long-term effects of conscription: Lessons from the UK*. Working Paper No. 0604. Department of Economics, University of Bergamo. Retrieved from http://goo.gl/EzxlV4.

Card, D., & Cardoso, A. R. (2012). Can compulsory military service raise civilian wages? Evidence from the peacetime draft in Portugal. *American Economic Journal: Applied Economics*, *4*(4), 57–93.

Grenet, J., Hart, R. A., & Roberts, J. E. (2011). Above and beyond the call: Long-term real earnings effects of British male military conscription in the post-war years. *Labour Economics*, *18*(2), 194–204.

Hahn, J., Todd, P., & van der Klaauw, W. (2001). Identification and estimation of treatment effects with a regression-discontinuity design. *Econometrica*, *69*(1), 201–209.

Harmon, C. P., Oosterbeek, H., & Walker, I. (2003). The returns to education: Microeconomics. *Journal of Economic Surveys*, *17*(2), 115–156.

Holland, P. W. (1986). Statistics and causal inference. *Journal of the American Statistical Association*, *81*(396), 945–960.

Imbens, G. W., & Angrist, J. D. (1994). Identification and estimation of local average treatment effects. *Econometrica*, *62*(2), 467–475.

Imbens, G. W., & van der Klaauw, W. (1995). Evaluating the cost of conscription in the Netherlands. *Journal of Business & Economic Statistics*, *13*(2), 207–215.

Imbens, G. W., & Wooldridge, J. M. (2009). Recent developments in the econometrics of program evaluation. *Journal of Economic Literature*, *47*(1), 5–86.

Keller, K., Poutvaara, P., & Wagener, A. (2009). Military draft and economic growth in OECD countries. *Defence and Peace Economics*, *20*(5), 373–393.

Keller, K., Poutvaara, P., & Wagener, A. (2010). Does a military draft discourage enrollment in higher education? *FinanzArchiv*, *66*(2), 97–120.

Lau, M. I., Poutvaara, P., & Wagener, A. (2004). Dynamic costs of the draft. *German Economic Review*, *5*(4), 381–406.

Lindo, J. M., & Stoecker, C. (2014). Drawn into violence: Evidence on "what makes a criminal" from the Vietnam draft lotteries. *Economic Inquiry*, *52*(1), 239–258.

Malig, J. (2011). WikiLeaks: US sees Filipino migrant workers as social "safety valve". abs-cbnNEWS.com, Retrieved from http://goo.gl/lBWHz

McCrone, P. R., Knapp, M. R. J., & Cawkill, P. (2003). Posttraumatic stress disorder (PTSD) in the armed forces: Health economic considerations. *Journal of Traumatic Stress*, *16*(5), 519–522.

Oi, W. Y. (1967). The economic cost of the draft. *American Economic Review*, *57*(2), 39–62.

Paloyo, A. R. (2010). *Compulsory military service in Germany revisited*. Ruhr Economic Paper No. 206. Retrieved from http://www.rwi-essen.de/publikationen/ruhr-economic-papers/304/

Poutvaara, P., & Wagener, A. (2007). To draft or not to draft? Inefficiency, generational incidence, and political economy of military conscription. *European Journal of Political Economy*, *23*(4), 975–987.

Reyes, V. (2011). *PAF mulls longer service contracts for its pilots*. Malaya. Retrieved from http://goo.gl/BzCRL

Schleicher, M. (1996). *Die Ökonomie der Wehrpflicht: Eine Analyse unter besonderer Berücksichtigung der Grundsätze der Besteuerung*. Frankfurt am Main, Germany: Peter Lang Verlagsgruppe.

Smith, J. P. (1999). Healthy bodies and thick wallets: The dual relation between health and economic status. *Journal of Economic Perspectives*, *13*(2), 145–166.

Spencer, D. L., & Woroniak, A. (1969). Valuing transfer of military-acquired skills to civilian employment. *Kyklos*, *22*(3), 467–492.

Strauss, J., & Thomas, D. (1998). Health, nutrition, and economic development. *Journal of Economic Literature*, *36*(2), 766–817.

US Department of Veteran Affairs. (2011). Retrieved from http://goo.gl/KxZm7

Wald, A. (1940). The fitting of straight lines if both variables are subject to error. *Annals of Mathematical Statistics*, *11*(3), 284–300.

REGIONAL REPRESENTATION IN THE U.S. MILITARY

Cindy Williams

ABSTRACT

The chapter examines regional differences in U.S. military participation. Participation in the U.S. all-volunteer military is persistently higher for youth from the South and West than from other regions. While raw numbers of recruits from the South and West have grown, much of the growth reflects broad demographic trends across the United States. In explaining broad differences among states as well as changes over time, the most important factors appear to be the prevalence of veterans in the state and regional differences in civilian pay levels. Remarkably, some characteristics that can disqualify individuals for military service are more pronounced in states that contribute more than their "fair share" of service members. Regional differences in military participation can have important implications for the relationship between armed forces and society. They also have consequences for communities: those that send more young people into the military may experience a disproportionate share of deaths and injuries during wartime; those that send fewer are less likely to enjoy the economic and training benefits associated with service.

The Evolving Boundaries of Defence: An Assessment of Recent Shifts in Defence Activities
Contributions to Conflict Management, Peace Economics and Development, Volume 23, 177–199

ISSN: 1572-8323/doi:10.1108/S1572-832320140000023012

A favorite Hollywood perception of the U.S. military is that it speaks with a Southern accent. To a significant extent, the perception is correct. In 2009, a young person from the Southeast of the country was about 70 percent more likely to join the armed forces than one from the Northeast. Military recruits are also more likely to come from rural areas than their civilian counterparts.

Regional and community differences in propensity to join the military are not new, and the wide variation in military participation across regions appears to feed on itself. The more a region sends to the military, the more veterans it generates; and areas with high concentrations of veterans generally contribute more recruits than those with fewer veterans. The services may also allocate more recruiters to locations where youth are more likely to say yes, and thus concentrate effort in areas of high propensity.

Regional differences in military participation can have important implications for the relationship between the armed forces and the society they serve. They also have important consequences for communities and regions: those that send more young people into the military may experience a disproportionate share of deaths and injuries during wartime; those that send fewer are less likely to enjoy the economic and training benefits associated with service.

This chapter discusses the geographic complexion of the U.S. military. It considers its causes and briefly discusses its consequences for the armed forces as well as for the regions and communities from which recruits are drawn.

BACKGROUND

Since its inception in 1973, the U.S. all-volunteer force has enjoyed substantial participation from the South. Southern representation in successive recruit cohorts grew significantly during the 1980s and 1990s. In recent years, Southerners made up more than 40 percent of new enlisted recruits across the services. The share of recruits who come from the West of the country has also grown, while the Northeast and North Central regions have lost ground (see Fig. 1).

To an important extent, the shift in recruits from North to South and from East to West reflects broad demographic changes of recent decades, as youth populations in the South and West grew more rapidly than those in the Northeast or Midwest. The Department of Defence takes account of such shifts in a measure that it calls the "representation ratio:" the fraction

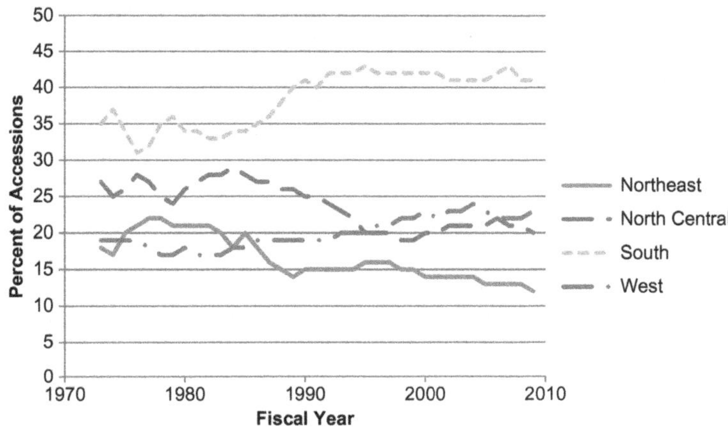

Fig. 1. Active Component Enlisted Accessions. *Source*: Author's graphic based on U.S. Department of Defense (DoD), Population Representation in the Military Services FY 2009.

Table 1. Representation Ratio for Enlisted Recruits, by Region.

Fiscal Year 2009	
Northeast	0.71
North Central	0.94
South	1.19
West	0.99

Source: DoD Population Representation in the Military Services, FY 2009, Table B-46.

of recruits coming from a region, divided by the fraction of 18- to 24-year-olds living in that region.[1] Tables 1 and 2 indicate the representation ratio for active-duty enlisted recruits in each of four broad regions of the United States, and then by state, during fiscal year 2009.[2]

With a representation ratio of 1.19 for the South, a young Southern man or woman was about 68 percent more likely to join the active-duty enlisted force of one of the services in 2009 than his or her counterpart in the Northeast, where the representation ratio is just 0.71. Montana contributed the greatest share of its age-eligible population; the District of Columbia was the lowest relative contributor, with a representation ratio of just 0.32. On average, a young person in Montana was about 5 times as likely to join the force as one from Washington, DC.

Table 2. Representation Ratio for Enlisted Recruits by State, FY 2009.

State	Representation Ratio	State	Representation Ratio
Alabama	1.23	Montana	1.57
Alaska	1.37	Nebraska	1.03
Arizona	1.28	Nevada	1.35
Arkansas	1.05	New Hampshire	1.06
California	0.82	New Jersey	0.65
Colorado	1.18	New Mexico	1.04
Connecticut	0.66	New York	0.66
Delaware	0.86	North Carolina	1.21
District of Columbia	0.32	North Dakota	0.65
Florida	1.38	Ohio	1.05
Georgia	1.34	Oklahoma	1.07
Hawaii	1.09	Oregon	1.35
Idaho	1.34	Pennsylvania	0.82
Illinois	0.79	Rhode Island	0.68
Indiana	1.06	South Carolina	1.28
Iowa	0.93	South Dakota	0.91
Kansas	1.01	Tennessee	1.13
Kentucky	0.92	Texas	1.25
Louisiana	1	Utah	0.73
Maine	1.32	Vermont	0.61
Maryland	0.82	Virginia	1.31
Massachusetts	0.59	Washington	1.12
Michigan	0.95	West Virginia	0.91
Minnesota	0.78	Wisconsin	0.9
Mississippi	0.96	Wyoming	1.04
Missouri	1.15		

Source: DoD, Population Representation in the Military Services, FY 2009, Table B-46.

Map 1 offers a snapshot of representation ratios by state in 2009. The broad middle of the country contributed a bit more than its "fair share," while the Northeast generally contributed less than its fair share. States in the Southeast as well as Texas and the West (except for California and Utah) contributed proportionately more.

The change in representation ratios from year to year is not as striking as the rise of Southern and Western participation in recruit cohorts when viewed in raw numbers. Fig. 2 shows representation ratios during selected years for nine U.S. regions: New England and the Mid-Atlantic (which together comprise the Northeast); East North Central and West North Central (which together make up the North Central area, formerly called

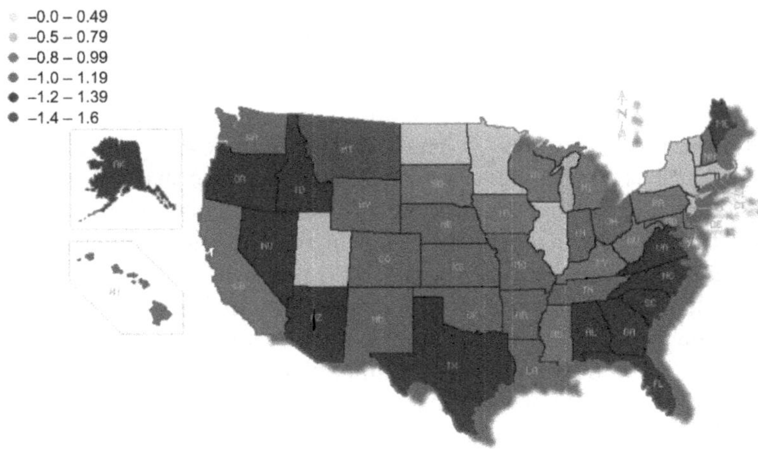

Map 1. Enlisted Recruit Representation Ratio. *Source*: Map created using software from DIYMaps.net; data from DoD.

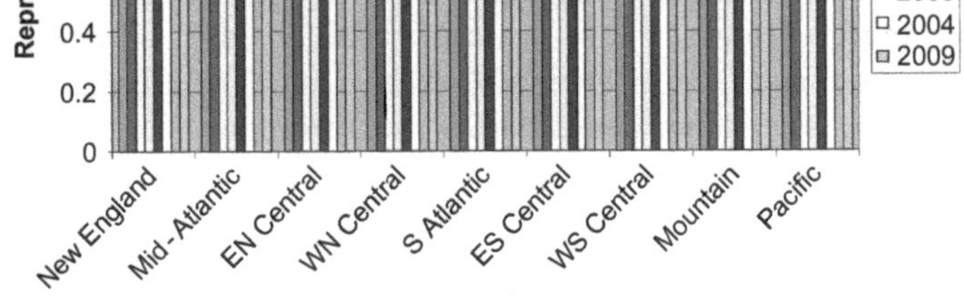

Fig. 2. Representation Ratio by Region for Selected Years. *Source*: Author's calculations based on DoD, Census Bureau, and Naval Postgraduate school.

the Midwest in U.S. Census reports); South Atlantic, East South Central, and West South Central (the South); and Mountain and Pacific (the West).

As the chart reveals, New England's representation ratio declined markedly between 1977 and 1989. Though recruiting from the region improved during the 1990s, it still sends a smaller share of its young people into the military than it did during the early years of the all-volunteer force. The share of young people joining the military from the West South Central region rose sharply between 1984 and 1989, and was still substantially higher in 2009 than in 1977. The Pacific region lost significant ground during the 1980s and has never returned to sending the share of young people it sent during the first years of the all-volunteer force (AVF).[3]

Variations in regional representation in the U.S. military are the result of the individual decisions of millions of Americans who make the choice, whether deliberately or not, of whether or not to serve their country in uniform. Each individual typically makes that choice for a complex mix of reasons. Some of those reasons are intangible, for example, patriotism, a desire to serve in an important institution, or a wish to help those fighting for freedom in other countries. Such intangible factors may well vary by region. For example, Samuel Huntington saw Southern participation in the military as reflecting traditions of romanticized violence, a Southern tradition of warfare, and a Southern agrarian culture.[4] Others see a Southern culture of manly honor and traditions of dueling as clues to the greater propensity of Southerners to join the force.

This chapter focuses largely on the tangible factors, which include demographic and economic conditions as well as the cognitive, educational, and physical qualifications of the young people whom the services might attract. Tangible factors also include policy choices made by the services' recruiting communities and, in some years, regional variations in recruitment incentives that the services instituted in order to assess the value of those incentives in attracting potential recruits. The next four sections discuss such tangible factors. Those sections are followed by a brief discussion of the possible consequences stemming from regional differences in representation ratios. The chapter ends with a short summary.

ECONOMIC FACTORS

Over the 38-year course of the U.S. all-volunteer force, the single most important driver of recruitment success or challenge has been the economy. Recruiters have to work harder when the economy is strong and unemployment is low, because young people finishing high school have plentiful opportunities for employment outside the military. Economic good times typically leave a mark on recruitment figures: numerical recruitment targets are less likely to be met, and the overall quality of the cohort is likely to be lower than during lean economic times with higher unemployment rates.

Figs. 3 and 4 show the U.S. unemployment rate and the quality of active-duty enlisted recruits from 1973 to 2009. Unemployment rates can

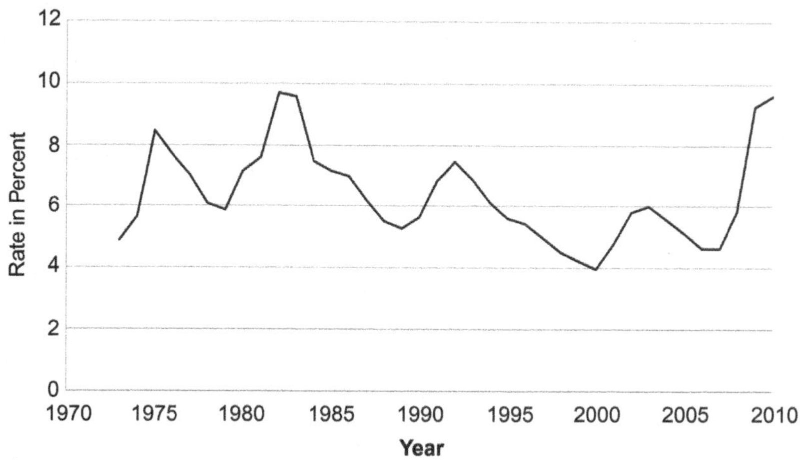

Fig. 3. U.S. Unemployment Rate 16 Years and Over, Annual Average. *Source*: Author's graphic, based on U.S. Bureau of Labor Statistics data.

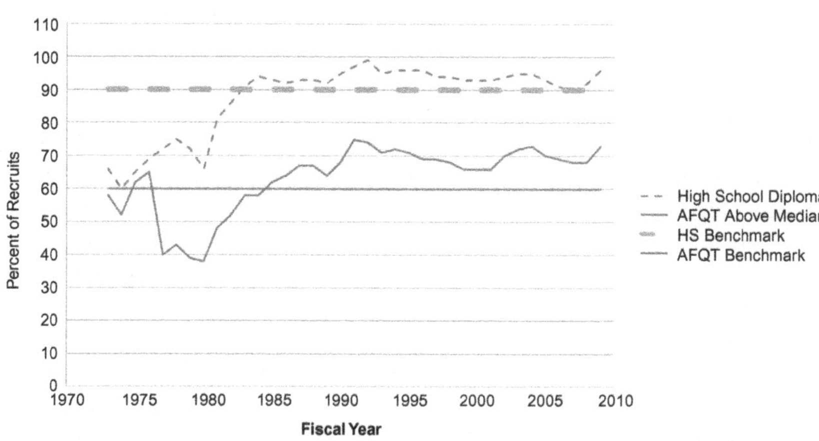

Fig. 4. U.S. DoD Recruit Quality Active Component Non-Prior Enlisted. *Source*: Author's graphic based on U.S. Undersecretary of Defense (Personnel and Readiness). *Note*: AFQT: Armed Forces Qualification Test Score; HS: High School.

vary significantly by state, and the economic factors that lead to high unemployment can affect individual states in varied ways. For example, the economic downturn of the late 1970s and early 1980s was particularly harsh in New England, the Mid-Atlantic, and the East North Central regions. That differential effect partly explains why the enlisted recruit ratios shot up in those regions during those years, only to decline again as the economies in those regions improved during the late 1980s.[5] When energy prices tumbled during the mid-1980s, the West South Central region, which includes four important oil and gas states – Texas, Oklahoma, Louisiana, and Arkansas – went from an oil boom to an oil bust practically overnight. The results are reflected in the "recruit boom" from that region in 1989.[6]

More recently, California and the Pacific region were especially hard hit, both by the bursting of the dot-com bubble in 2000 and from the financial and economic meltdowns of 2008 and 2009. Partly as a result, the share of recruits coming from the Pacific region rose.

Average wages for civilian workers also vary by state, making military pay seem more attractive to high-school graduates in some states than in others. In 2009, average annual wages were $37,000 or less in Arkansas, Iowa, Mississippi, Montana, North Dakota, South Carolina, South Dakota, and West Virginia. With the exception of North Dakota, those states had representation ratios greater than 0.9. In contrast, Alaska, California, Washington, DC, Maryland, Massachusetts, New Jersey, New York, and Washington had average wages above $47,000. Except for Alaska and Washington, those states contributed substantially less than their "fair shares" to the 2009 recruit cohort. Fig. 5 illustrates the negative correspondence between wages and military representation ratios in 2009.

Finally, there is a significant "rural versus urban" divide when it comes to participation in the U.S. military. In 2009, more than 38 percent of the Defence Department's new active-duty enlisted recruits came from rural areas.[7] By contrast, only 27 percent of Americans lived in those areas. Rural populations are generally more prevalent in the South than in the Northeast or the West.[8]

At least since the beginning of the all-volunteer force, rural areas have sent proportionally more recruits to the military than urban areas or suburbs. For example, in 1987, about 30 percent of new enlistees to the active or reserve forces came from rural counties, when just 25 percent of Americans lived in those counties.[9] (Differences in U.S. Census Department measurement categories may explain some of the differences between these figures and those for 2009.) The rural–urban divide is even

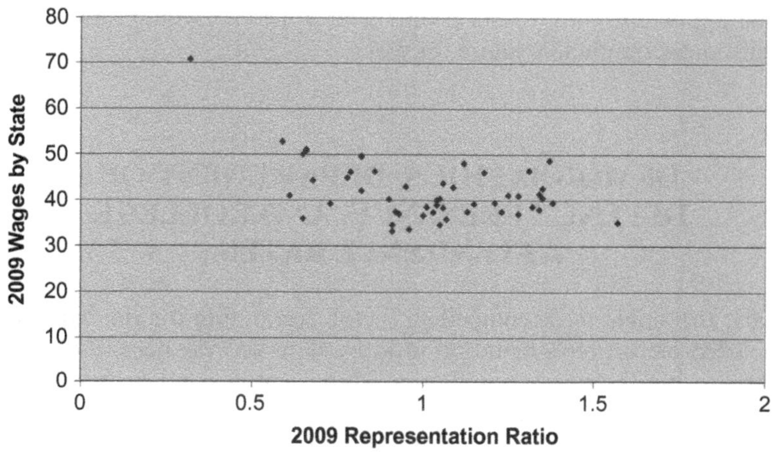

Fig. 5. Civilian Wages and Representation Ratio 2009. *Source*: Author's graphic based on U.S. Bureau of Labor Statistics and Department of Defense.

more pronounced for the National Guard than it is in the active-duty forces.[10]

The differences in military participation rates between rural and urban youth may explain why some states look like outliers in their regions when it comes to sending young people into the armed services. For example, Maine – with less industry and more rural territory than other New England states – sends a much greater share of its youth into the armed forces than the states that surround it. In contrast, California, anchored by two major metropolitan areas (San Francisco in the north and Los Angeles in the south), has a noticeably lower military participation rate than its close neighbors with less urban build-up.

A number of theories have been advanced as to why young people who grow up in rural areas gravitate more readily to military service than city dwellers. One is that children in rural areas are more likely to learn to hunt; thus taking up arms in the military comes more naturally for them. Another is that families in rural America are by nature more conservative, and therefore more likely to encourage their youth to enter service.

Social and cultural explanations may be part of the story, but earnings, economic opportunity, and the presence of military recruiters also play important roles. Wages are traditionally higher and job opportunities more plentiful in cities than in rural areas. In addition, the military stations recruiters across the nation; thus recruiters can become the local face of outside opportunity for young people who otherwise would have to travel long distances simply to explore for work.

DEMOGRAPHICS, DEPARTMENT OF DEFENCE PRESENCE, AND COLLEGE ATTENDANCE RATES

In 2009, the single most compelling factor accounting for the variation in the enlisted recruit representation ratio by state was the density of military veterans within the state's population. Map 2 shows the density of veterans by state. The maps of recruit representation ratio and veterans density are strikingly similar, as the scatter plot in Fig. 6 confirms.

The services' recruiting commands have long understood that the opinions of parents, grandparents, teachers, coaches, clergy, and other adult "influencers" can weigh heavily in a young person's decision to join the military. Veterans have personal experience of military life, and it seems that their presence in the population can have a profound effect on a young person's choice regarding military service. The opposite is also no doubt true: states that contribute proportionally more recruits to the military

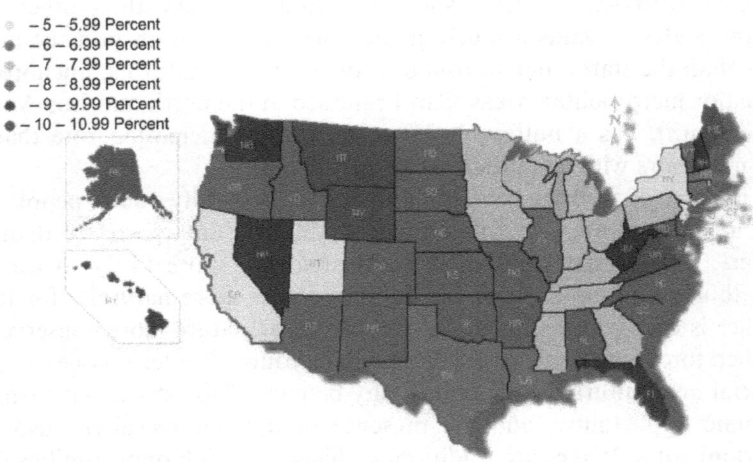

Map 2. Density of Veterans in the Population. *Source*: Map created using software from DIYMaps.net; data from U.S. Census Bureau.

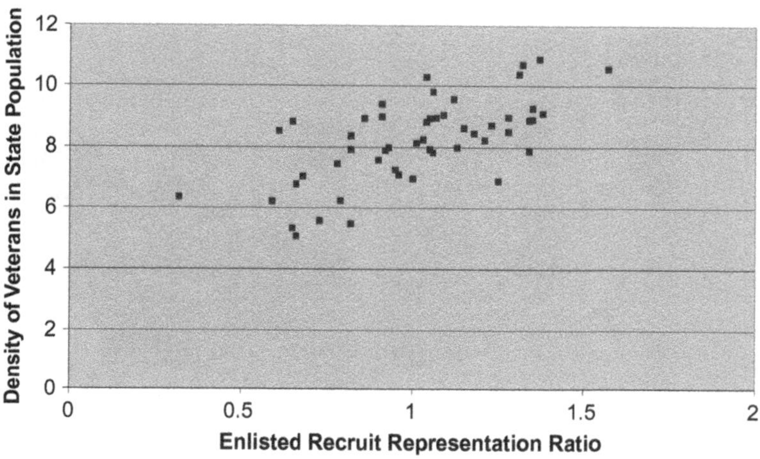

Fig. 6. Recruit Representation Ratio and Veterans 2009. *Source*: Author's graphic based on DoD and Statistical Abstract of the United States.

acquire more veterans, as many of those recruits return to their home states when they depart the service.

The presence of a sizeable active-duty military population in a state contributes only mildly to the share of the youth population that joins the military. Map 3 indicates the density of active-duty military personnel within each state's population. Alaska and Hawaii both host sizeable military installations, and both contribute far more than their "fair shares" of young people to the military. On the other hand, Maine, Florida, and Texas all contributed well above their fair shares of recruits in 2009, but have small or moderate representations of active-duty military personnel at military installations within their borders. This mixed picture is reflected in the scatter diagram of Fig. 7.[11] Any effect from military procurement within a state is also mild (see Map 4).[12]

State-wide college attendance rates have an important effect on recruitment. In states where more young people attend college after completing high school, fewer are available to join the military. Table 3 indicates college entry rates for high-school graduates by state in 2008.

Note for example that Massachusetts, New Jersey, and New York had particularly high continuation rates from high school on to college during 2008. Those states all had low enlisted recruit representation ratios in 2009. States with especially low college entry rates include Alaska, Idaho, and Oregon – all of which contributed well above their fair shares to the 2009

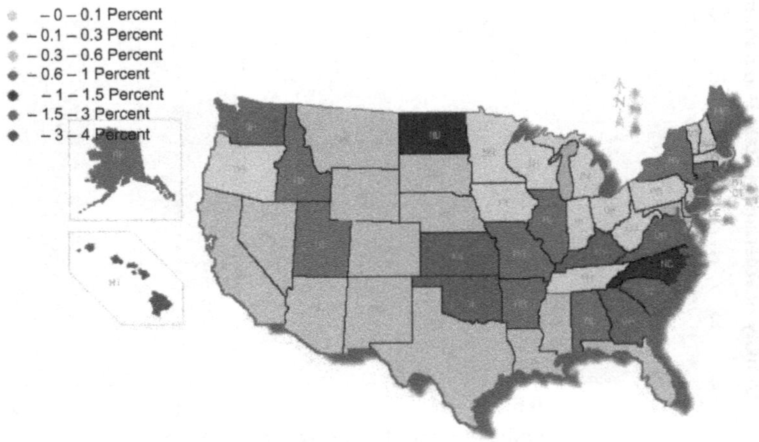

Map 3. Density of Active-Duty Military Personnel. *Source*: Map created using software from DIYMaps.net; data from U.S. Census Bureau.

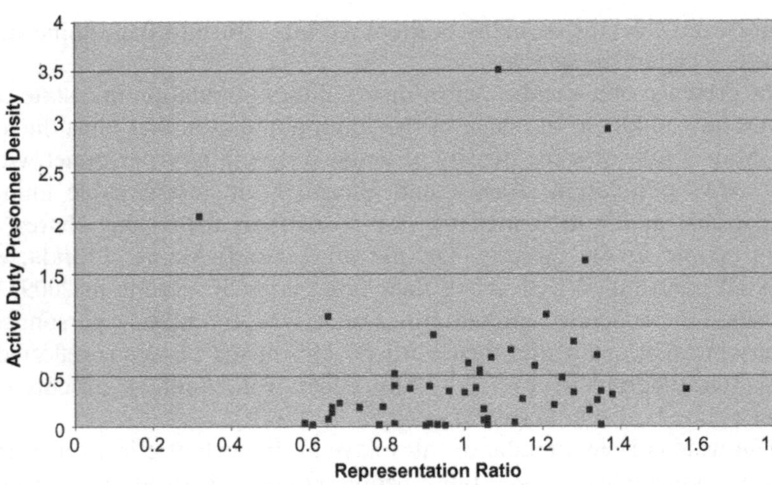

Fig. 7. Representation Ratio and Active-Duty Density 2009. *Source*: Author's graphic based on DoD and U.S. Census Bureau.

recruit cohort. The fit is not perfect, however. For example, more than 70 percent of South Carolina's high-school graduates continue directly on to college, yet that state sends substantially more young people into the active-duty enlisted force than most other states (adjusted for youth

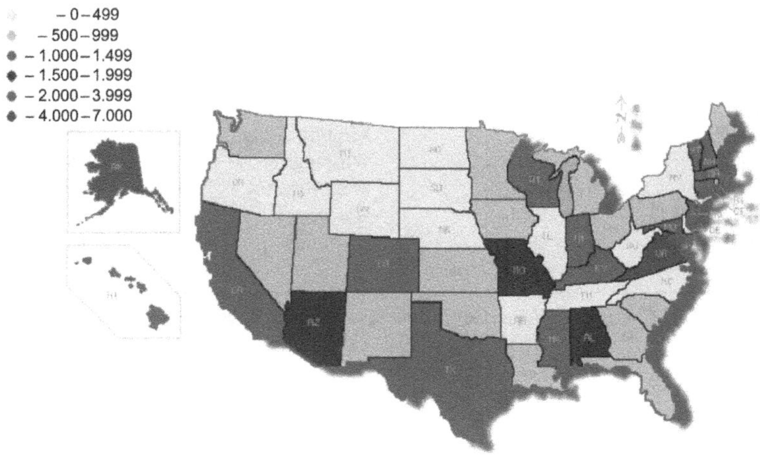

Map 4. Defense Procurement Spending per Capita. *Source*: Map created using software from DIYMaps.net; data from DoD.

Table 3. College Entry Rates of High-School Graduates, 2008.

State	Rate of College Entry (Percent)	State	Rate of College Entry (Percent)
Alabama	66.7	Montana	51.8
Alaska	45.7	Nebraska	65.5
Arizona	51.4	Nevada	55.6
Arkansas	62.5	New Hampshire	63.9
California	65.4	New Jersey	71.1
Colorado	62.6	New Mexico	67.7
Connecticut	68	New York	74.2
Delaware	66.2	North Carolina	66
Florida	58.8	North Dakota	67.6
Georgia	69.6	Ohio	62.7
Hawaii	62.3	Oklahoma	56
Idaho	49.1	Oregon	46.5
Illinois	57.4	Pennsylvania	63.9
Indiana	65.7	Rhode Island	67.4
Iowa	64.3	South Carolina	70.1
Kansas	65.4	South Dakota	72.1
Kentucky	60.9	Tennessee	61.6
Louisiana	65.3	Texas	56.9
Maine	61.3	Utah	58.5
Maryland	62.9	Vermont	48.3

Table 3. (*Continued*)

State	Rate of College Entry (Percent)	State	Rate of College Entry (Percent)
Massachusetts	74.6	Virginia	68.7
Michigan	59.9	Washington	50.7
Minnesota	69.2	West Virginia	59
Mississippi	77.4	Wisconsin	59.1
Missouri	60	Wyoming	59.4
Montana	51.8		

Source: National Center for Higher Education Management Systems.

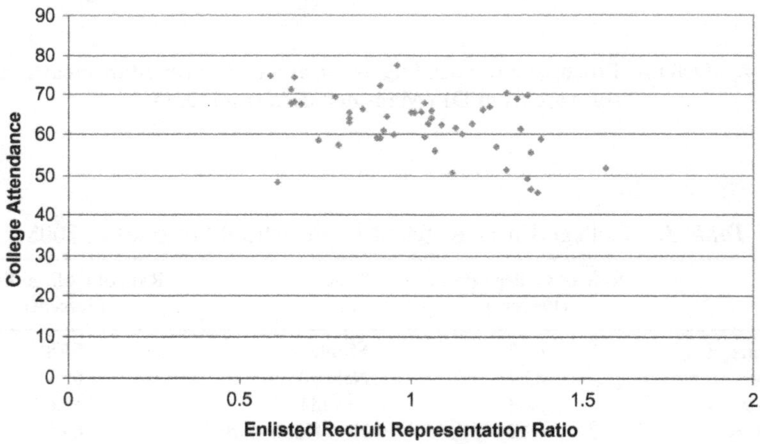

Fig. 8. Rate of College Attendance after High School, 2008. *Source*: Author's graphic based on DoD and National Center for Higher Education Management Systems.

population). Similarly, fewer than one-half of Vermont's high-school graduates go on to college, yet Vermont is one of the lowest providers of military recruits relative to its youth population. Fig. 8 maps the negative relationship between states' rates of college attendance and their enlisted recruit representation ratios. Not every young person who wants to join the military is eligible for service. The following section discusses the role that differences in eligibility can play in regional representation.

VARIATIONS IN ELIGIBILITY FOR MILITARY SERVICE

In 2007, the Lewin Group estimated that only 4.7 million of the 31.2 million Americans between 17 and 24 years old were qualified for military service. Some 35 percent of Americans in that age group are medically or physically unfit – in many cases because they are overweight. Another 18 percent are disqualified for prescription or non-prescription drug use. Others do not meet services' expectations for high-school graduation or cognitive aptitude.[13]

Such factors vary significantly by state and region. Regional recruitment figures reflect not just the differences in individuals' desire to join the military, but these differences in eligibility as well.

The United States today is in the grip of an obesity epidemic. More than 23 percent of American adults between the ages of 18 and 34 have a body-mass index (BMI) of more than 30, which corresponds to an individual who is 5 feet 4 inches (1.62 meters) tall and overweight by more than 30 pounds (13.6 kilograms). The rate of obesity is about 2.5 times as high as it was in 1991.

In 1991, high rates of obesity were concentrated in a few states of the South and Midwest. Today, every state has an obesity rate greater than 15 percent, with the highest concentrations in the South and Midwest (see Map 5). For states with higher concentrations of obesity, it is harder for young people to join the military, because a large fraction of them are disqualified by virtue of their weight. The South and Midwest might thus contribute an even larger share of their youth to military recruit cohorts if those youth could meet the services' weight standards.

Another important disqualifier from military service is the failure to complete high school with a high-school diploma.[14] Department of Defence quality benchmarks call for at least 90 percent of new enlisted entrants to be high-school diploma graduates.[15] High-school graduation rates vary sharply by state, with rates as high as 93 percent for Iowa and as low as 54 percent for Georgia. Like obesity, the failure to graduate from high school appears to be more prevalent in the South than in the Northeast. Thus the South's military representation ratio might be even higher if more of its young people graduated from high school. (On the other hand, if better high-school graduation rates inspired more young people to attend college, it is possible that the effect could work in the opposite direction.)

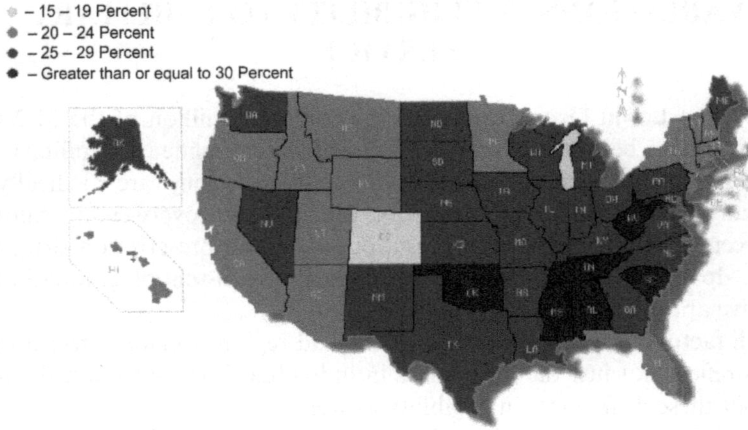

Map 5. Obesity Trends among U.S. Adults, 2008. *Note*: Obesity indicates BMI greater than or equal to 30, about 30 pounds overweight for 5-foot 4-inch person. *Source*: Map created using software from DIYMaps.net; data and obesity definition from Behavioral Risk Factor Surveillance System, U.S. Center for Disease Control and Prevention (CDC).

Finally, the services set benchmarks for performance on their entrance test, the Armed Forces Qualification Test (AFQT). The AFQT measures cognitive aptitude, and is somewhat similar to an intelligence test. Traditionally, young men and women from the Northeast, North Central, and Western regions perform better on the AFQT than those raised in the South. This means that proportionally more Southerners are disqualified based on the AFQT score than those from other regions. On the other hand, better test performance and high-school graduation rates can make New England good territory for recruiters despite the region's persistent low representation ratios.

POLICY CHOICES AND EXPERIMENTS

U.S. military recruitment specialists often point out that the military is not just an "all-volunteer force;" rather it is an "all-recruited force." Indeed, a substantial share of the young people who join did not set out to do so as they considered their options during high school. Instead, they were recruited by the professionals each service sends to recruiting stations in communities across the country.

The services make deliberate choices each year about how many recruiters to station in each location, what goals to establish for each recruiter, and how to incentivize recruiters to meet their targets. As local labor markets change, they apportion the recruiting force and adjust the goals set for recruiters. Thus, if the jobless rate rises dramatically in a state, the Army might reduce the number of recruiters stationed there, knowing that enlistees will be easier to draw into the force. Alternatively, if jobs are plentiful and recruiting more difficult, the Air Force might add recruiters there, lower the number of individuals each recruiter is meant to bring in, and raise the incentives for recruiters who meet or exceed their goals.

On the other hand, if a community traditionally yields strong recruiting results, then the services typically send more recruiters there than to communities where fewer young men and women are likely to answer the call to service.[16] Like the presence of veterans in a community, this apportionment of effort tends to feed on itself: The more recruiters the service assigns, the more young people they recruit into the service. And the more successful the recruiters are, the more the service assigns to the community.

An additional factor sometimes complicates the recruiting picture: Department of Defence experiments aimed at identifying the relative effectiveness of a variety of incentives offered to young people who enlist. During the first three decades of the all-volunteer force, the services conducted several such experiments.[17] The experimental designs generally varied the incentives, such as college education or contract terms, by state. Thus for some years, the relative recruiting picture improved in some states and deteriorated in others as a result of deliberate design choices in the experiment.

THE CONSEQUENCES OF REGIONAL RECRUITING DIFFERENCES

For a variety of reasons, the U.S. military draws more on some states and regions than on others for its recruits, and draws more heavily on rural areas than on cities. This has been true for decades, for the officer corps as well as the enlisted force, and for the National Guard and Reserve as well as the active-duty force.[18]

Regional and community differences in military representation have implications for the armed forces and their relationship with the society they serve. For example, Army leaders may worry that as fewer New Englanders volunteer to serve, fewer future Senators and Congressmen from the region will have a close understanding of the Army. That may make it harder to persuade them of the legitimacy of Army budget requests.

More generally, because politics vary by region in the United States, the regional differences in military participation may contribute to a widening political and cultural gap between the nation's military and its civilian leaders, and ultimately to problems in civil–military relations.

A 1990s survey of officers and civilians found that military officers are significantly more conservative than civilian elites and also see themselves as being more conservative than the broader public views itself to be.[19] Officers are also much more likely to identify with the Republican Party than are civilians; and the share of the officer corps that identifies as Republican grew markedly between the mid-1970s and the late-1990s.[20]

Michael Desch argues that because the South is heavily represented in the U.S. officer corps, the officer corps was more profoundly affected by the rise in Republican Party identification that took hold in the South between the 1970s and the 1990s than were other segments in society.[21] As the South shifted to the Republican Party, it brought the heavily Southern officer corps along with it, putting military leaders increasingly at odds with civilian elites in their political and social views.[22]

Yet it is those elites whose decisions dictate what the nation will spend on its military, how it conducts its foreign affairs, and when and how the United States goes to war. Some observers worry that the growing political and cultural gap could ultimately lead to distrust and disrespect of the military toward the civilian government.[23] The concern also goes the other way: If civilians perceive service members as vastly different from themselves, they may withdraw their support and respect for the military as an institution.[24]

Regional differences also have important consequences for the regions and communities themselves. An obvious example is that regions that send more young people into the military are likely to experience a disproportionate share of deaths and injuries during wartime. At the height of the Iraq war, much was made of the disproportionate share of the burden carried by rural America. On the other hand, regions that contribute significantly to the military may benefit from the work skills, leadership capacity, and levels of maturity that returning service members might bring home after a period in the military.

In addition, if more of a region's lawmakers have military experience, they might gravitate more readily to the congressional committees with substantial influence over the services' modernization choices. Thus, those regions might ultimately find it easier to compete for military contracts.

For many communities, heavy participation in the Guard and Reserve can be a double-edged sword. A rural community whose population participates widely in the National Guard benefits economically from the influx of training pay during peacetime. On the other hand, that same community may lose an important component of its workforce when the training unit is deployed to an operation. In Alaska, for example, several National Guard units consist largely of native-American hunters and fishermen who are the main breadwinners for their communities. When such a unit is deployed, entire communities can lose a substantial fraction of their subsistence base.

Another issue related to heavy participation in the Guard and Reserve is that many reservists serve as police, firefighters, or emergency workers in their communities. When their units are called to active duty, their communities may not have enough emergency responders to deal with a natural disaster or other crisis at home.

CONCLUSION

At least since the beginning of the all-volunteer force, the U.S. military has drawn its recruits disproportionately from the South and West of the country. Every individual who chooses to join the armed services does so for his or her own mix of reasons. Differences in regional representation reflect an amalgam of complex choices regarding education, employment, the example and advice of influencers, and a variety of intangible factors by millions of young people every year.

Nevertheless, a few factors stand out as particularly important in determining both the persistent differences in military representation among states and the changes in representation ratios over the years. Prominent among the explanations is that states with relatively large populations of military veterans are considerably more likely to contribute above their "fair share" of youth to the armed services. Since many service members return to their home states upon departure from the military, the prevalence of veterans in a state is almost surely both a reason for greater military participation by the next generation and a result of greater participation by the previous generation.

Within a region, the unemployment rate can be an important factor in changes in military participation from year to year. When unemployment is high, military service can be an attractive alternative for those with lowered job prospects. Under conditions of low unemployment, employment opportunities in the private sector are plentiful, and fewer young people gravitate toward service in the military. Pay levels for civilian workers vary widely by region, and are also an important factor in choices to join the military. In the United States, military basic pay is the same regardless of the region from which an individual joins the service. Thus, military pay looks more attractive to a young person in a region where civilian pay is low than to one from a region where civilian pay is higher.

When it comes to military service, the United States faces a significant rural–urban divide, with significantly higher military participation rates for individuals who come from rural areas. The rural–urban divide explains some of the regional and state-by-state differences in military participation rates. That divide in turn is often attributed to cultural differences between rural and urban youth, but it also clearly reflects differences in economic and employment opportunity.

Military recruiters and recruitment policies also play an important role. Even individuals who are eager to join the military need a recruiter to open the door for them and guide them through the process. Moreover, the services rely heavily on attracting individuals who would otherwise not have chosen to serve in uniform. Thus, the number of recruiters assigned to a region and the goals and incentive structures established for them can make a big difference among states and across time periods.

Regional differences in military participation rates can have important consequences for civil-military relations. For example, the difference in military participation between North and South may be an important factor in the emergence of a widening political and cultural gap between the nation's military officer corps and its civilian leaders. Some fear that the growing gap may lead officers to distrust and disrespect the civilians charged with deciding what to spend on the military and when and how to go to war. Viewed in the other direction, if civilians perceive service members as vastly different from themselves, they may withdraw their support and respect for the military as an institution. That could translate into smaller military budgets as well as a lack of understanding and appreciation of the sacrifices made by men and women in uniform.

Regional differences in military participation also have implications for the regions and communities themselves. Rural America paid a significantly higher casualty price during the wars in Iraq and Afghanistan than did urban centers, because young men from rural areas are more likely to join the military than their urban or suburban counterparts. On the other hand, regions that contribute significantly to the military may benefit from the skills returning service members bring home. For the nation's citizen soldiers, heavy participation in the Guard and Reserve may bring economic benefits, but it can also mean long absences when key workers or first responders, such as police and firefighters, are called up for deployment.

NOTES

1. The U.S. armed services set limits on the age of first entry. Entrants as young as 17 years old are permitted. Currently, the Army is willing to consider new applicants who are as old as 41 years of age. Nevertheless, the preponderant age for non-prior-service enlisted recruits is between 18 and 24 years old. In 2009, 86 percent of new enlisted recruits across the Department of Defense fell in that age band. See U.S. Undersecretary of Defense (Personnel and Readiness) (2009), Table B-1.

2. Tables 1 and 2 as well as most of this chapter focus on active-duty, non-prior-service enlisted recruits – that is, those who join the active-duty enlisted force with no prior service in the military. The enlisted force comprises about 84 percent of the U.S. active-duty military, with officers and warrant officers making up the remaining 16 percent. Prior-service accessions are only a small fraction of all enlisted accessions. The Guard and Reserve contribute some 850,000 paid troops to the services, while the active components numbered about 1.5 million total uniformed personnel in 2011. Some of the regional and community differences discussed in this chapter are even more pronounced for the Guard and Reserve than they are for the active components.

3. Data for the figure are drawn from the following sources: Greenwood and Mehay (1991), Table 9: Distribution of DOD Non-Prior Service Accessions and Youth Population (Percent) (various years). The 1984 and 1989 figures include effects from recruiting experiments conducted during those years: the Enlistment Bonus Test conducted between July 1982 and June 1984 and the Army 2+2+4 experiment that ran from July 1989 to October 1990. See Rostker (2006).

4. Huntington (1957).

5. Greenwood and Mehay (1991, p. 26).

6. Greenwood and Mehay (1991, p. 27).

7. U.S. Undersecretary of Defense (Personnel and Readiness) (2009), Table B-42; U.S. Census Bureau, Statistical Abstract of the United States: 2011, Table 28.

8. Race no doubt also plays a role, not examined in this study. Except in the South, American Blacks are generally underrepresented in rural areas of the country. Compared with their share of the overall population, they make up a larger share of urban populations in the Northeast and Midwest. Patterns are changing, however. For example, the decade of 2000 witnessed a significant migration of Black workers to cities in the South. Except at the height of the war in Iraq between

2004 and 2009, the share of Blacks in annual cohorts of military recruits was significantly higher than the share in the population of 18- to 24-year olds in the nation.

9. Fernandez (1989).

10. Fernandez (1989, 2009).

11. Large military bases may actually result in twin effects that offset each other. On the one hand, the presence of a base can lead more young people to become familiar with the military. On the other hand, military bases typically spell employment for civilian government workers and contractors, thus reducing economic incentives for young people to consider joining the military.

12. Like military installations, military shipyards and factories can familiarize young people with military weapons and serve to attract them to join up; on the other hand, they provide good jobs for high-school educated workers and thus reduce the incentive to join the service.

13. OUSD (P&R)/Military Personnel Policy presentation, October 28, 2010, slide 14.

14. The services draw a sharp distinction between graduation by passing an equivalency test and graduation with a diploma, because staying in school until graduation with a diploma is a strong indicator that an individual will stay in the military for the full contracted term of service.

15. In the United States, those who do not graduate from high school can earn a General Equivalency Diploma (GED) by passing a battery of tests in five subject areas. The Department of Defense draws a distinction between graduating from high school with a diploma and earning the GED, and the GED is generally considered insufficient for entrance into the military. Historical evidence indicates that enlisted members who are high school diploma graduates are more likely to complete their service contracts than GED holders.

16. Greenwood and Mehay (1991, p. 27).

17. Rostker (2006). Experiments discussed by Rostker include the Multiple Option Recruiting Experiment (MORE) of 1979, p. 475; the Navy Enlistment Marketing Experiment of 1979, pp. 478–479; the Enlistment Bonus Test, which ran from July 1982 to June 1984, p. 614; and the Army 2+2+4 Experiment, from July 1989 to October 1990, p. 634.

18. As regards the officer corps, see Desch (2001). For the reserve component, see Fernandez (1989).

19. Feaver and Kohn (2001).

20. Feaver and Kohn (2001). It should be noted that the military-civilian gap is a phenomenon of the officer corps and generally not apparent for the enlisted force. See Segal, Freedman-Doan, Bachman, and O'Malley (2001).

21. Desch (2001). This chapter focuses on regional representation in the enlisted force, but the regional differences it highlights are even more pronounced in the officer corps.

22. Desch (2001).

23. In fact, the military officers surveyed in the Feaver and Kohn study "criticized the quality of civilian leadership and expressed a pervasive hostility toward the media." On the other hand, both officers and enlisted personnel were more likely to express "trust and confidence in government institutions" than do civilians. Feaver and Kohn (2001, p. 461).

24. This has not happened yet, however. Like many surveys in recent years, the most recent Gallup Poll on the subject, June 23, 2011, found that Americans have greater confidence in the military than in any other institution in America.

REFERENCES

Desch, M. C. (2001). Explaining the gap: Vietnam, the republicanization of the South, and the end of the mass army. In P. D. Feaver & R. H. Kohn (Eds.), *Soldiers and civilians: The civil-military gap and American National Security* (pp. 289–324). Cambridge, MA: The MIT Press.

Distribution of DOD Non-Prior Service Accessions and Youth Population (Percent). (Various Years). *Population representation in the military services.* Washington, DC: U.S. Department of Defense, various years.

Feaver, P. D., & Kohn, R. H. (2001). Conclusion. In P. D. Feaver & R. H. Kohn (Eds.), *Soldiers and civilians* (pp. 459–473). Cambridge, MA: The MIT Press.

Fernandez, R. (1989, October). *Social representation in the U.S. Military* (pp. 1–2). Washington, DC: Congressional Budget Office.

Fernandez, R. (2009). *Population representation in the military services.* Washington, DC: U.S. Department of Defense.

Greenwood, M. J., & Mehay, S. L. (1991, March). *Trends in regional patterns of migration, immigration, and economic activity: Implications for army recruiting.* Monterey, CA: Naval Postgraduate School.

Huntington, S. P. (1957). *The soldier and the state: The theory and politics of civil-military relations* (pp. 211–212). Cambridge, MA: Harvard University Press.

Rostker, B. (2006). *I want you: The evolution of the all-volunteer force* (pp. 614–634). Santa Monica, CA: Rand.

Segal, D. R., Freedman-Doan, P., Bachman, J. G., & O'Malley, P. M. (2001). Attitudes of entry-level enlisted personnel: Pro-military and politically mainstreamed. In P. D. Feaver & R. H. Kohn (Eds.), *Soldiers and civilians* (pp. 163–212). Cambridge, MA: The MIT Press.

U.S. Undersecretary of Defense (Personnel and Readiness). (2009). *Population Representation in the Military Services* 2009. Washington, DC: U.S. Department of Defense.

WOMEN AND CONFLICTS, THE ADAPTATION OF UN 1325 RESOLUTION IN THE EUROPEAN UNION

Steven Coissard and Liliane Perrin-Bensahel

ABSTRACT

This chapter aims to present the implementation of the United Nations 1325 resolution about the rights of women, including their protection against violence. This resolution is the main international instrument which incorporates and mandates a gender perspective in all aspects of peace building, from prevention to conflict resolution. It is part of a necessary evolution of gender relations in peace building since 1975 and the Conference of Mexico. More broadly, this text would lead to improve the women's representation presence in the social and economic sphere and their important role in development.

The Evolving Boundaries of Defence: An Assessment of Recent Shifts in Defence Activities
Contributions to Conflict Management, Peace Economics and Development, Volume 23, 201–218

ISSN: 1572-8323/doi:10.1108/S1572-832320140000023013

Gender issue has long been neglected in the context of war and conflicts, even though research is developing for few decades. This failure is especially true at the international level, since one has to wait until the beginning of the 21st century before the Security Council of United Nations seriously took the status and place of women into consideration. Since October 2000, a legal framework has been indeed defined.

Resolution 1325 (2000) and the following are the result of a long process of mobilisation that has taken over 20 years (from Mexico in 1975 to Beijing in 1995). Obstinacy and several mass gatherings were needed to finally get the recognition of the dramatic situation of women in conflict areas. In October 2009, the Security Council was still concerned about obstacles that prevent women from a full protection during a conflict or from contributing to the prevention and resolution of conflicts. Such concerns lead all involved parties to organise activities to raise awareness of resolution 1325 at national, regional and international levels and to press states for implementing this resolution.

The UN 1325 resolution was a significant step and landmark in the protection of women and children. It was the first time that the Security Council of the United Nations was interested in the experiences of women and their specific contribution to peace and security. Resolution 1325 calls for full participation of women in respect of gender and mainstreaming gender in all initiatives relating to peace and security. It covers the closely interrelated areas of participation, protection, prevention, support and repair. The decision came while the prosecution for rape and violence against women were a significant number of cases handled by the International Tribunal.

In the first part, we examine the approach of gender and conflict, then resolution 1325 and its context and finally its application in Europe and France.

THE GENDER APPROACH AND CONFLICTS

NGOs of women using gender approach have invested heavily in conflict with the support of the United Nations and therefore many studies have been conducted to support implementation south field, like those conducted by the laboratory Bridge in Britain (http://www.bridge.ids.ac.uk/).

The starting point of the gender approach is the fact that men and women have different positions in society, different life experiences, different expectations in their professional and private life and different values. These differences, which assign to men and women different roles, are social constructs that arise from history, are related to culture and are not determined. Gender refers thus to the social and cultural identity of a man or a woman determined by an assignment of duties, functions and roles of each sex in society, in public life and private life. This is a cultural definition of femininity and masculinity and also a social definition about the relations (hierarchy) between men and women.

The gender approach also allows the recognition of women role in the society:

- The recognition of civic rights and citizenship of women close to those of men;
- A contribution to the participation of the civil society in strategies and regional, national and international policies;
- Their important participation in the development of the society (economic activities, educational activities, traditional and religious aspects);
- The interaction between men and women (factor of division of labour, allocation of resources and time);
- Studies concerning gender relate to the social construction of men and women, that is, to the roles and expectations determined by the same society that shapes men's and women's life;
- Finally, the four social roles of women: reproductive role (child care, housework, health, supply), productive role (income generation), community role (activities and volunteer contributions to the social and community life) and political role (exercise of power and participation in decision-making at all levels).

This raises the following question: is the introduction of gender approach could destabilise and shape our institutions, organisation and/or society? Academics tried to understand and enhance the role of women in society and define two main concepts: the mainstreaming and the empowerment.

The mainstreaming consists of two approaches: the integrated approach, related to the notion of equal opportunity, and the participative approach, aiming to awake on men/women relationship. Current expressions of 'mainstreaming' are the principle of positive discrimination and parity. However, this concept raises many questions and the necessary means for its implementation are weak. For instance, despite the adoption of the Charter for equality between Men and Women in 2004, France has restricted the scope of the concept. In the *Journal Officiel* of 4 March, 2006, was issued a notice of the General Commission of Terminology and Neologisms defining the concept of 'equal representation' as an 'action for equality between women and men' and as an equivalent of gender mainstreaming. Representation and action are two different terms with different level of implication.

Gender mainstreaming was endorsed in 1995 at the 4th World Conference on Women organised by the United Nations in Beijing. Since the ratification of the Amsterdam Treaty in 1997, it is enroled in the missions of the European Union (Article 3). Gender mainstreaming takes into account the principle of equality between women and men in all phases, at all levels and for all areas of decision-making. As such, it is part of the process of strategies and policies development. Mainstreaming involves ensuring that gender perspectives and attention to the goal of gender equality are central to all activities: policy development, research, advocacy/ dialogue, legislation, resource allocation and planning, implementation and monitoring of programmes or projects. It calls on governments and other stakeholders to encourage the adoption of active and visible to ensure the inclusion of gender in all policies and programs in order to analyse the impact on men and women, respectively, before making any decision.

In its resolution of 21 September 1995, the European Parliament considers that it is essential to continue to actively coordinate and integrate policies for equity and equal opportunities and to consider any politics, programs structures in a gender perspective (Official Journal of the European Communities, 1995). Mainstreaming integrates the specific characteristics of gender in the development and implementation of policies and economic and social measures, but also the analysis of consequences that these measures could have on men and women (Labourie-Racapé, 1997).

Besides, the Gender Empowerment Measure is one of the five indicators used by the United Nations Development Programme in its annual *Human Development Report* (since 1995). It is a composite index measuring gender inequality in three basic dimensions of empowerment-economic participation and decision-making, political participation and decision-making and power over economic resources. These dimensions consider first the parliamentary representation with the seats held by women in parliament, second the female legislators, seniors officials, managers and their professional and technical positions, and finally the power over economic resources estimated in their earned income.

Moreover, according to Le Bossé and Lavallée (1993) empowerment includes five essential components: participation, competency, self-esteem and critical awareness (individual, group, social and political awareness). It restores the idea that women and men have an aptitude that can be acquired to make choices. The various levels of intervention of empowerment concern: well-being (satisfaction of women practical needs); equality in access to resources and services; awareness (conscientious understanding of the difference between sex and gender, collective participation of women); participation (equal participation of women in all decision-making process); control of decision-making, wealth and benefits (Marius-Gnanou, 2007).

Women are discriminated against in many countries. Women represent 67% of illiterate, 10% of revenues, 1% of the property. Sixty-seven per cent of hours worked in the world are allocated to women. Nutrition and food security, the role of women is crucial, especially in food production (e.g. 1990, Rwanda: 79% of working hours devoted to food production are performed by women; Togo: 57% of farmers are women, '40% of ploughing, planting 80%, 70% weeding and harvesting, and almost any vegetable production on-farm are provided by women').

Approaches to gender and training women who have been involved have a significant mobilisation through, including NGOs. Dissemination of information, increasingly frequent discrimination against women and services they face, particularly in conflict situations, have to hear the mobilisation of NGOs and the need for a text to protect women and children. The resolution 1325 is the result of both awareness and mobilisation. This shows both their role in the development of societies but also their role as guardian of the family and their commitment to peace.

WOMEN AND CHILDREN: MAIN VICTIMS OF THE WAR

'Feminist theory teaches us conflicts, wars and militarism are gendered processes. They use, maintain and often promote the ideological construction of gender in the definitions of "masculinity" and "femininity" and of course have disproportional impact on women and children, particularly the girl child. In effect, it means that men go to war to defend national/state values, territories and borders and to protect and defend their "own" women and children. Women are seen as "the protected" and "the defended," which inevitably means women having to "survive the violence" and "patch and mend the war-torn societies" instead of their equal participation in contributing to the democratic development, enforcement of rights and justice, and creating human security for all' (European Women's Lobby, 2009).

Seventy per cent of victims of recent conflicts are women and children. The female body is a battlefield; women are raped, abducted, humiliated and reduced to sexual slavery. In this case, women are victims on three levels: as woman, as guardian of the family and as member of an ethnic group. Moreover, after becoming a victim of conflicts, they are very often excluded of their own community. There are several experiences of such situation in many countries: Afghanistan, Burundi, Colombia, Ivory Coast, Liberia, Peru, Republic of Congo, Sierra Leone, Darfur, and former Yugoslavia. Half of men indicted by the International Court of Justice are accused of rape or sexual assault.

Eighty per cent of rapes in the context of armed conflict are collective and largely public. Between 300,000 and 500,000 children, one-third of girls, are enroled in armies in over 20 countries. From 1990 to 2000, teenage girls were involved in armed conflicts in at least 39 countries. Three quarters of displaced persons living in refugee camps are women and children. In Rwanda, half a million women were raped during the genocide of 1994; 20,000 children would be born of such acts. 60,000 in Croatia and Bosnia and Herzegovina; 64,000 in Sierra Leone between 1991 and 2001 (Adequations, 2008). Rape is a strong symbol of victory. Women become the 'stake' and the 'place' of the battle or conflict. Systematic rape is a weapon of terror, a weapon of war, a strategy psychologically used to eliminate resistance or an instrument of torture used in special prisons out of sight. They are an instrument of war and ethnic cleansing, similar to genocide.

During World War II, 200,000 women from Korea and Indonesia have been removed for use 'of comfort women' for Japanese army; the Wehrmacht organised mass and public rapes (Stienne, 2011); the French army or the American forces were not exempted of such atrocity. Rapes are often justified by ethnic cleansing. In former Yugoslavia, 'rape camps' were open for the procreation of the 'species'. Today, these women are still living close to their aggressors and are afraid to speak.

The costs of violence against women are extremely high. They include direct costs of care and support to abused women and their children and the costs of bringing to justice the perpetrators of violence. Indirect costs include loss of worked hours and productivity as well as pain and human suffering.

UN 1325 RESOLUTION AND ITS CONTEXT

Resolution 1325

Resolution 1325 is the culmination of a long and laborious process conducted jointly by the UN and women's organisations engaged in the

'Program for Action the Fourth World Conference on Women' to mainstream a gender perspective into peacekeeping operations. It covers the closely interrelated areas of participation, protection, prevention, support and repair. It aims to call on all parties involved in an armed conflict to take special measures to protect women and girls from gender violence, including rapes and other forms of sexual abuse. Resolution 1325 consists of 18 items classified in four main recommendations: representation of women, training on gender issues, fair treatment and respect for international law.

As the weight of women has significantly increased in all operations to promote and maintain peace, including local initiatives, Resolution 1325 highlights the importance of education to military and civilian personnel in charge of peacekeeping issues concerning the protection, rights and needs of women and an awareness of HIV/AIDS (Article 6). Furthermore, in order to protect women and girls, it makes governments and non-state actors responsible for crimes committed during conflicts. To sum up:

- The resolution aims to a better representation of women, training on gender issues, fairness of treatment and compliance to international laws.
- It recognises that it is important to take into account the gender approach, and women specific needs, at all levels of conflict management and peace development.
- Women's power has considerably increased in all missions aiming to promote and maintain peace, including local activities.

Resolution 1325 is the main international instrument that incorporates and mandates a gender perspective in all aspects of peace building, from prevention to conflict resolution. It means that a perspective of gender equality must be implemented throughout the peace process in the treaties/agreements and peacekeeping process.

Resolution 1325 stresses four main areas of concern:

- The increased representation of women at all levels of decision-making and participation in peace building and conflict prevention;
- A gender perspective in post-conflict processes (including peace processes) and training to promote gender perspective in peace building;
- An attention to specific protection needs of women – including from sexual violence;
- A gender perspective in the UN programming, reporting and peace operations reports.

Six strategic objectives were set for actions by governments, intergovernmental, regional and international institutions:

- First objective: Widening women's participation to conflict resolution at the decision-making level and protect women living in situations of armed conflicts;
- Second objective: Reduce excessive military expenditures and control arm stocks;
- Third objective: Promote non-violent forms of conflict resolution and reduce violations of fundamental women's rights in conflict situations;
- Fourth objective: Promote women's contribution to the development of a culture of peace;
- Fifth objective: Give protection, aid and training to refugees, to other displaced women in need of international protection and to women displaced within their own country;
- Sixth objective: Assist women from former colonies and non-autonomous territories.

To support Resolution 1325, several resolutions were taken.

Resolution 1820 (2008) was the first to deal exclusively with the sexual violence in armed conflict. The Security Council agreed that sexual violence is a security issue and noted that the use of violence as a tactic of war against civilian populations is a threat for the maintenance of peace and security. 'Rape and other forms of sexual violence can constitute war crimes, crimes against humanity or a constitutive act with respect to genocide.' On the national level, the resolution urged states to ratify or accede to the Convention on the Elimination of all forms of Discrimination Against Women (CEDAW), the Convention on the Rights of the Child, and the Optional Protocols for each of them.

Resolution 1888 (2009) stressed the importance to include questions relating to sexual violence in the early process of peace and to bring to justice theirs perpetrators. It called for specific provisions for protection of women and children in the mandates of the UN peacekeeping operations and of peace negotiations sponsored by the Organisation. The Security Council referred to the need for taking into account the issue of violence sex at an early stage of the peace process and efforts mediation, to protect people risk and promote a full stabilisation, particularly in the areas of agreements on access for humanitarian stakeholders before the ceasefire and on human rights, ceasefire disarmament, demobilisation and rehabilitation, and the reform of security sector.

Resolution 1889 (2009) reaffirmed the Security Council's determination to ensure women rights in its different resolutions. It leaded to the nomination of a special representative of the UN General Secretary in charge of the fight against sexual violence in conflicts (Miss Margot Wallström) and to the consolidation of information about women's situation in conflicts. Resolution 1889 called for measures providing tools for the Security Council to set up a monitoring system. The resolution mandated the production of a set of indicators on Resolution 1325 for use at the global level, which were produced in March 2010 in an inter-agency process led by United Nations Development Fund for Women (UNIFEM).[1]

Resolution 1960 (December 2010) asks to the parties engaged in armed conflicts to organise sanctions for men who develop acts of violence against women.

These different resolutions are completed by institutional mechanisms (the Special Representative of the United Nations Secretary General on Sexual Violence in Conflict) and relevant international instruments: CEDAW (Convention sur l'Elimination de toutes les formes de Discriminations à l'Egard des Femmes (CEDEF)/CEDAW, 1979), the platform for action of the Beijing Conference (1995), the Rome Statute of the International Criminal Court (1998) and the Millennium Development Goals (especially MDG 3 on gender equality and on autonomy, 2000).

UN Resolution 1325 in Europe and France

The stock of resolutions, ten years after the creation of Resolution 1325, is still hesitant. Indeed, up to now, only a handful of countries have developed national action plans (NAPs). Ten EU countries have already prepared it: Austria (2007), Belgium (2009), Denmark (2007), Finland (2008), the Netherlands (2009), Portugal (2009), Spain (2007), Sweden (2006), the United Kingdom (2006), as three countries outside EU: Iceland, Switzerland and Norway. Canada, Ivory Coast, Liberia and Uganda are the only non-European countries to have developed an NAP. Even if such plans exist, this is barely sufficient. The related funding process and implementation are not always formalised.

UN Resolution 1325 and the European Union

During the French Presidency of the EU in 2008, several studies have been conducted in this field. The French Presidency of the European Union (July–December 2008) was commissioned to construct indicators for the critical area no. 5 'women and armed conflict'. It placed at the heart of priorities the need to work on indicators for this issue to create ways for its implementation and to fight against impunity for war crimes and crimes against humanity.

During this period, some decisions were taken:

- The fight against violence on women and the promotion of women's role in conflict resolution and post-conflict reconstruction became the priority of human rights.
- The adoption of EU guidelines on violence against women and the fight against all forms of discrimination against them.
- The formalisation of a comprehensive approach to the European Union for the implementation of resolutions 1325 and 1820.

The EU established the components of its plan to be transposed later in the different European countries. The Treaty of Lisbon will affect some aspects of the EU's policy on armed conflict in some important ways for women. In particular, it extends the obligation of gender mainstreaming to the EU's security policies. The creation of a European External Action Service (EEAS) bringing together structures and staff from the European Commission, the Council Secretariat and Member State services offers a unique opportunity to ensure the equal representation of women and men in this area, as well as the full implementation of gender mainstreaming in new policy-making procedures. Catherine Ashton became the head of this new service as High Representative of the Union for Foreign Affairs and Security Policy.

The civil society made 10 proposals for a better implementation of Resolution 1325 and commitments taken by European countries in terms of Women Peace and Security (WPS):

1. Facilitate and enhance participation of women in peace and security as a priority.
2. Include a set of minimum standards in all plans/strategies of action relating to the WPS.
3. Involving civil society organisations in the development, implementation, monitoring and evaluation of plans for the WPS.
4. Include meaningful indicators and mechanisms for monitoring and evaluation plans/strategies.
5. Allocate human and financial resources specific to the WPS.
6. Appoint a senior representative for matters regarding WPS.

7. Ensure that the EEAS contributes more to the implementation of Resolution 1325 and related resolutions, and works in accordance with commitments related to the WPS.
8. Strengthen the Working Group of the WPS of the European Union for better coordination and implementation.
9. Ensure compliance with commitments related to the missions of the Common Security and Defence Policy in WPS.
10. Prepare an annual report on the implementation at the European level of commitments related to the WPS.

For Ines Alberdi, 'It is time for us all to count the number of women at the peace table, the number of women raped in war, the number of internally displaced women who never recover their property, the number of women human rights defenders killed for speaking out. All of this counts, and we are counting'.[2] The absence of women and their perspectives in peace negotiations, post-conflict reconstruction, disarmament, humanitarian relief and peace building are a persistent and direct barrier to the integration of human security and women rights' concerns and a barrier to sustainable peace.

UN 1325 Resolution and France

In France, the Ministry of Foreign and European Affairs has set up a NAP. This plan calls for inter-ministerial and inter-institutional coordination. The action plan is structured in four parts (Ministère des Affaires Etrangères Européen, 2010).

First part: Protecting women against violence and working to ensure respect for their fundamental rights

- Ensuring that the personal situation of women victims of an armed conflict is taken into account in the procedure for gaining access to international protection outside their country of origin, on the basis of relevant standards;
- Providing specific assistance to women victims of violence (physical, sexual and psychological) during conflict and post-conflict periods as part of humanitarian aid programmes, in cooperation with all competent European and international stakeholders, including access to health care – including sexual and reproductive one – and psychological support;
- To ensure the protection of women against all forms of violence during post-conflict periods, through reinforced coordination of actions and resources allocated as part of development cooperation

to respect for the rights of women and girls and the promotion of gender equality.

Second part: Participation of women in managing conflict and post-conflict situations

- Ensuring support for the participation of women in the decision-making and political processes relating to the reconstruction of countries affected by international or domestic armed conflict or post-crisis situations, including processes relating to traditional justice, international criminal justice and the struggle against impunity;
- Ensuring support for civil society organisations working for gender equality, empowerment of women and strengthening of women's equal participation in all decision-making processes;
- Encouraging reinforced direct participation of women in peace-keeping missions and reconstruction operations by encouraging their access to responsibilities within civilian and military components and to high-ranking positions in the chain of command.

Third part: Raising awareness of respect for women's rights in training programmes

- Including issues related to respect for women's rights and protection against all forms of violence in training related to peacekeeping and security;
- Systematically including a better awareness of the respect for women and girls' rights and gender equality in actions related to security system reform, peacekeeping and security and support for transitional justice processes;
- Providing technical and financial support to encourage French speaking countries wishing to develop expertise in this field to do so, in partnership with the 'International Organisation of La Francophonie' and research/training centres;
- Fostering education in non-violence, human rights and peace culture in schools through a multidisciplinary approach and specific trainings.

Fourth part: Developing political and diplomatic actions

- Promoting the adoption of civilian protection mandates in peace-keeping operations so as to improve prevention, fight violence against women and protect victims;
- Systematically including women's rights and gender equality in France's bilateral and regional dialogues on security issues (particularly with the African Union and the other regional organisations);

- Contributing to the development and implementation of European guidelines: European Union guidelines on violence against women (evaluation and implementation of regional strategies and dialogue with third countries), monitoring of implementation indicators of the Beijing Action Platform on 'Women and Armed Conflict' and the European Union's 'Women, Peace and Security' indicators;
- Supporting, in compliance with competence criteria, the nomination and appointment by the United Nations Secretary General of women as representatives and special envoys in charge of missions and as Human Rights specialists employed in the field (as part of the United Nations and the European Union operations);
- Supporting the European Union's fight against impunity and the policy of tolerance by maintaining active support for international criminal justice (and especially the work of the International Criminal Court).

Concrete commitments and indicators, timetable and list of concerned ministry complete these objectives.

The Possible Effects of This Application in France
It is now time to ask questions about these evolutions. The time of the French Presidency of Europe was important to work on the protection of women in conflict and led to a growing awareness of the issue and the implementation of the resolution 1325 in France. The presentation of the implementation of resolution 1325 by France is by no means a model but a testimony. The introduction of gender approach could be very destabilising and giving large transformation in the organisation, culture of an institution or a society. This approach queries the question of time, of social links and representation on different places and role of men and women.

French Context. The implementation of Resolution 1325 in France depends on several ministries and not just the ministry of defence. It also takes into account the situation of violence against women in countries in peace like France. In this country every third day a woman dies under the blows of his spouse or companion. Moreover, women's rights are violated at work, where with equal qualifications and responsibilities, they are paid less, where the 'glass ceiling' is present and does not always allow access to positions of highest responsibility, where uncertainty affecting more women than men. A comprehensive approach to this issue is not only important but also necessary if we want to change the ratio of sexes in society and make the recognition of the role of women wealth and a basis for development. This is the position taken by the UN and claimed by France even if the application is not always obvious.

In France, the reform of compulsory military service and the professionalisation of the armed forces have led to major changes including an acceleration of the integration of women. The French armed forces are well in the first place in Europe for their feminisation with 14% of women, not far from the United States (15.5%). Although it began in the 1970s, this evolution is not achieved. Indeed, some specialties remained banned for women. For example, they were not allowed to board warships until 1993. In 1998, quotas were abolished with the implementation of the principle of gender equality in access to various military corps. If the presence of women has increased in armed forces, it becomes scarce when you go up the hierarchy (10% for officers, 5% for senior officers). Several factors explain this situation: due to the recent access to military jobs for women and the length of contracts, their number is still low in military schools; moreover, women do not stay long in armed forces (leaving on average after 15 years of career, while men usually leave after 25–30 years of career).

The feminisation of the French army is linked to the professionalisation but also to new types of conflicts, the changing role of the army as in the prevention of situations of war, and the evolution of the requested technical professionals.

The report from the Economic and Social Council on 'The role of women in the profession of arms' in 2004 reported on the perception of the contribution of women in armed forces by military leaders. It recognised an important contribution to replace conscripts. On the qualitative level, girls and women have a higher educational level than their male colleagues. Their presence seems to facilitate certain tasks in police-related missions (especially in the Gendarmerie) such as spinning, the information in the investigation of police to make contact with the inner-city populations, juvenile protection. They are better on managing prevention measures, and perseverance, organisation, pragmatism, seriousness and intuition are feminine qualities recognised and appreciated.

Possible Impacts. The image of the armed forces is very manly, in line with its traditional role. However, many missions have evolved and prevention becomes an important issue, especially concerning women and children. The female presence among peacekeeping intervention forces is therefore useful. Control of women in conflict zones as Afghanistan is very important, and men cannot do this task.

Women's participation in officer corps would evolve strategic choices and strategy. For example, their sensitivity with regard to particular situations of women would lead to a better prevention and care of populations. It is also important that abused women could talk with other women and permit effective mutual recognition and mutual confidence. Information will be more easily brought to Court and the guilty punished. Stopping rapes as an act of war should be a goal worn by men and women managers in a modern army.

Moreover, the position of women about war seems different from men. The work of NGOs bringing together women from countries in conflict such as Jerusalem Link (coordination of two independent women's centres: Bat Shalom – The Jerusalem Women's Action Center, located in West Jerusalem, and Marcaz al-Quds la l-Nissah – The Jerusalem Center for Women, located in East Jerusalem) or Women for Peace are significant. We can also note the experience of women in Burundi who have created a network, the CAFOB (collaboration of association and women's NGOs in Burundi) which has two objectives:

(1) Developing dialogue and solidarity between women of different ethnic groups to have the same vision of conflict in Burundi and the same approaches to peace building;
(2) Supporting the solidarity and cohesion through income generating activities. They aim to preserve the conditions of peace on both sides with the same concerns: a condition of life, survival, protection of children and family (Ndacayisaba, 2001).

Generally, we say that economy is neutral but it's a male point of view. The introduction of gender is opening other possibilities and especially in the defence and security sectors. The new French law in reference to the Resolution 1325 should change the organisation, culture and approach of the French defence institutions. Consequences will be, among others, a reform of security sector, an evolution and recognition of women's role in development and an increase of number of women in politics. In France, women hold 44.4% of parliamentarian seats in the European Parliament in 2009, 21.8% in the Senate in 2008, 18.9% are deputies, ranking France in 18th place in European Union. Women are better represented in regional council (48% in 2010) but they represent only 13.8% of the elect in general councils (at *département* level) and 13.9% in city councils. There are 20.2% of women leaders in the public sector while they are 59.3% of civil servants (rank A). 9.9% of Prefects are women, 6.5% of treasurers-paying, 11% of ambassadors .... They represent less than one-third of management positions in the private and semi-public companies. The percentage of women in the boards of CAC 40 companies was 15.3% in 2010.[3]

The participation of women in development began to be recognised. Their presence in negotiations for resolution of conflict and reconstruction are the underpinnings of a lasting peace. Their desire to preserve the living conditions can influence some decisions during the negotiations. On the other hand, increasing their presence in companies' boards or as members of Parliament and senators should improve the recognition of their participation in economic and social development, the need to respect their rights, roles and weight in the decision process on strategies and policies to implement.

In the world, women continue to have a lower representation in formal peace processes, although they contribute in many informal ways to conflict resolution. In recent peace negotiation, women have represented fewer than 8% of participants and fewer than 3% of signatories.

CONCLUSION

The recognition and implementation of Resolution 1325 are of course only the beginning of a process that needs to be developed, expanded and deepened. This movement is essential to change gender relations. Now, not respecting the women, their rights and their place in the society is on the one hand to forego an extraordinary richness that is half the world population and on the other hand it is for men to forego substantial support and opportunities to change our civilisation and all the men and women who compose it. Women can also release the men to free themselves of the injunctions that society assigns to them and so live the life they want to live.

NOTES

1. UNIFEM is women's fund at the United Nations. UNIFEM was created by a UN General Assembly resolution in 1976.
2. "10 points on 10 years of UNSCR 1325", October 2009, Civil society recommendations in the Implementation of UNSCR 1325 in Europe, http://expert.care.at/downloads/careexpert/CS_Recommendations_1325inEurope.pdf.
3. Service des droits des femmes et de l'égalité de la Direction générale de la cohésion sociale, 2010.

REFERENCES

Papers

Adequations. (2008). *Les femmes et les conflits armés*. Rapport de la présidence française de l'UE, Décembre. Retrieved from http://www.adequations.org. Accessed on December 13, 2011.

Artioli, F. (2013) The navy and the city: Conflict, cooperation and political competition in the urban governance of Toulon. *Urban Research & Practice*, *6*(1), 75–94.

European Women's Lobby. (2009). Position paper on gender and conflict toward human security: Engendering peace', October 30. Retrieved from http://www.womenlobby.org/spip.php?%20article93&lang=en. Accessed on December 13, 2011.

Harding, A. (1991). The rise of urban growth coalitions, UK-style? *Environment and Planning C: Government and Policy*. *9*(3), 295–317.

Hohenberg, P., & Lees, L. H. (1995). *The making of urban Europe, 1000–1994*. Cambridge, MA: Harvard University Press.

John, P. (2001). *Local governance in western Europe*. London: Sage.

John, P., & Cole, A. (1999). Political leadership in the new urban governance: Britain and France compared. *Local Government Studies*, *25*(4), 98.

Labourie-Racapé, A. (1997). Oser la parité pour l'égalité. In Serge Cordellier et Elisabeth Poisson (coord.), *L'État de la France 97–98*, Paris, La Découverte.

Le Bossé, Y., & Lavallée, M. (1993). Empowerment et psychologie communautaire: Aperçu historique et perspectives d'avenir. *Cahiers Internationaux de Psychologie Sociale*, *20*, 7–20.

Marius-Gnanou, K. (2007). Mobilités de travail et reconstruction des rapports de genre. Séminaire interne, Pacte-IGA/PEPSE-ESE, Grenoble, 15 Janvier.

Ndacayisaba, M.-G. (2001). Impact d'un conflit sur les relations hommes-femmes: Cas du Burundi. In *Hommes armés, femmes aguerries: Rapports de genre en situations de conflit armé*/textes réunis par Fenneke Reysoo (pp. 117–125). Genève: IUED.

OECD. (2006). *OECD territorial reviews: Competitive cities in the global economy*. Paris: Organization for Economic Co-operation and Development (OECD).

Provincia di Taranto. (2012). Comune di Taranto, Autorità Portuale di Taranto, Confindustria Taranto, Camera di Commercio di Taranto, et CGIL CISL UIL Taranto. *Protagonismo Locale, Sostegno Regionale e Nazionale per una Stagione di Risarcimento al Servizio del Risanamento e dello Sviluppo dell'Area Ionica*. Taranto: Provincia di Taranto.

Sellers, J. M. (2002). *Governing from below: Urban regions and the global economy*. Cambridge: Cambridge University Press.

Stienne, A. (2011). Viols en temps de guerre, le silence et l'impunité. *Le monde diplomatique – Blog*, *14* Février. Available at http://blog.mondediplo.net/2011-02-14-Viols-en-temps-de-guerre-le-silence-et-l-impunite. Retrieved on 3 June 2014.

Official Reports/Documents

Ministère des Affaires Etrangères Européen. (2010). National Action Plan for France – Implementation of the "Women, Peace and Security" resolutions of the United Nations Security Council. Retrieved from http://www.diplomatie.gouv.fr/fr/IMG/pdf/PNA_en_DEF.pdf

Official Journal of the European Communities. (1995). *Résolution du Parlement européen du 21 septembre 1995 sur la Quatrième Conférence mondiale de la Femme à Pékin: Égalité, développement et paix*, n°95/C 269.

Service des droits des femmes et de l'égalité de la Direction générale de la cohésion sociale. (2010). *Les chiffres-clés de l'égalité entre les femmes et les hommes.*

PART IV
THE INTERNATIONAL LANDSCAPE OF DEFENCE

DEFENCE EXPENDITURES: THEORY AND EMPIRICS

Carlos Seiglie, Scott Yi-Chun Lin and Tanu Kohli

ABSTRACT

There has been an extensive amount of research on the determinants of military spending over the last 25 years. These studies underline that military spending is a complicated concept, with economic capabilities, political processes and military linkages playing an interdependent role at the national, regional and global levels. Recent works focus on other outcomes of military spending. This chapter develops a model of conflict that generates a demand for military personnel and equipment by countries for either aggressive or defensive purposes. This model highlights some of the key determinants of military spending. Using pooled time-series, cross-sectional data on military spending for 146 countries from 1998 to 2007 we test this model and analyze other possible factors that previously have not been explored in the literature.

The Evolving Boundaries of Defence: An Assessment of Recent Shifts in Defence Activities
Contributions to Conflict Management, Peace Economics and Development, Volume 23, 221–230

ISSN: 1572-8323/doi:10.1108/S1572-832320140000023014

There has been an extensive amount of research on the determinants of military spending over the last 25 years. These studies include Griffin, Wallace, and Devine (1982), Maizels and Nissanke (1986), Seiglie (1988, 1992), Seiglie and Liu (2002) and Smith (1980, 1989, 1990) amongst others. For example, Maizels and Nissanke (1986) conclude that military spending is a complicated concept, with economic capabilities, political processes and military linkages playing an interdependent role at the national, regional and global levels. Seiglie (1988, 1992) develops a model with a utility maximising individual that values national security and from it generates a derived demand for military spending which depends upon several factors. Similarly, Smith (1989) uses the linear expenditure system to determine the demand for military expenditure by a country contingent upon the prices, strategic factors and on the military spending of other countries.

More recent works of Gupta, de Mello, and Sharan (2001) and Dunne, Nikolaidou, and Mylonidis (2003) focus on other outcomes of military spending. For example, Dunne et al. (2003) examine the change in military spending during 1980s and 1990s and conclude that while international conflicts have given way to internal conflicts, there have been small changes in military expenditures. Gupta, de Mello, and Sharan (2001), on the other hand, look at corruption and military spending and place a positive relationship between them such that countries with greater military spending also have higher chances of governmental corruption. This chapter develops a model of conflict that generates a demand for military personnel and equipment by countries for either aggressive or defensive purposes. This model highlights some of the key determinants of military spending. We then empirically test the model by using pooled time-series, cross-sectional data on military spending for 146 countries from 1998 to 2007. In addition, we explore the impact of other factors on defence spending which have not been previously explored in the literature.

THE BASIC MODEL

We begin by assuming two countries which are denoted by *A* (Attacker) and *D* (Defender) who are in dispute over claims to a territory or some other resource which is initially part of *Country D*. They both place a common value of *V* on this disputed territory. *Country A*, which is the attacker wants to take possession of this disputed territory. We assume that there are two periods. In the first period, Country *D* and Country *A* decide their respective military expenditure or stock of weapons which we denote by F_D and F_A, respectively. In the second period, they both enter into a conflict over the territory. If Country A wins, they capture the territory from Country D. Conversely, if Country D wins they retain the territory. In addition, if either country loses the war, they experience an additional loss which we denote by L_A and L_D, respectively. This can be interpreted as casualties from war, resources expended during the conflict or a decline in reputation internationally.

Following Grossman and Kim (1995, 1996), the probability of winning (or the proportion of the territory retained by the defender) depends upon the relative amount of defence versus attack effort. More specifically, the percentage retained of their endowment, r, or contest-success function is:

$$r = \frac{F_D}{F_D + F_A} \tag{1}$$

Therefore, the defender retains (wins with a probability) r and the attacker retains $(1 - r)$ of the value of the defender's endowment (V) during an attack where:

$$(1 - r) = \frac{F_A}{F_D + F_A} \tag{2}$$

Each country has preferences for consumption, c_{it} in each of the two periods represented by the following utility functions:

$$U_D = c_{Dt} + \beta_D c_{Dt+1} \tag{3}$$

$$U_A = c_{At} + \beta_A c_{At+1} \tag{4}$$

In addition, each country is initially endowed with E_D and E_A of undisputed resources which they can consume or use for weapons in any period. The budget constraints for the defender in each period are:

$$c_{Dt} + F_D = E_D + V \tag{5}$$

$$c_{Dt+1} = r(E_D + V) + (1 - r)(E_D - L_D) \tag{6}$$

Eq. (5) states that the defender's consumption plus military expenditures in period t cannot exceed its endowment or income generated by its undisputed and disputed resources, E and V. Similarly, Eq. (6) specifies that consumption in the second period, $t+1$, must be equal to their expected endowment in that period where if they win they keep the proportion r of the disputed territory and if they lose they have $(E_D - L_D)$ remaining.

The budget constraints for the attacker in each period are:

$$c_{At} + F_A = E_A \tag{7}$$

$$c_{At+1} = (1-r)(E_A + V) + r(E_A - L_A) \tag{8}$$

Substituting Eqs. (1), (2), (5) and (6) into Eq. (3) yields the defending country's expected utility function:

$$U_D = (E_D + V - F_D) + \beta_D \left(\frac{F_D}{F_D + F_A}(E_D + V) + \frac{F_A}{F_D + F_A}(E_D - L_D) \right) \tag{9}$$

Similarly, the attacker country's expected utility function is:

$$U_A = (E_A - F_A) + \beta_A \left(\frac{F_A}{F_D + F_A}(E_A + V) + \frac{F_D}{F_D + F_A}(E_A - L_A) \right) \tag{10}$$

The problem for each country is to maximise their expected utility by choosing their amounts of weapons. Therefore, maximising Eqs. (9) and (10) with respect to F_D and F_A, respectively, we get

$$F_D = \sqrt{\beta_D F_A (V + L_D)} - F_A \tag{11}$$

$$F_A = \sqrt{\beta_A F_D (V + L_A)} - F_D \tag{12}$$

Eq. (11) is the *best-response function* or the reaction function for the defender that establishes the relationship between the military spending of *Country D* and that of *Country A*. Similarly, Eq. (12) is the reaction function for the attacker. Solving for the Nash equilibrium (F^*_D, F^*_A) of this game yields that

$$F^*_D = \frac{\beta_A\beta_D^2(V+L_A)(V+L_D)^2}{[\beta_A(V+L_A)+\beta_D(V+L_D]} \tag{13}$$

$$F^*_A = \frac{\beta_D\beta_A^2(V+L_D)(V+L_A)^2}{[\beta_A(V+L_A)+\beta_D(V+L_D)]} \tag{14}$$

It is clear that from Eqs. (13) and (14) we can derive a series of comparative statics regarding the effects of changes in the parameters on military spending. Finally, Eqs. (11) and (12) represent a two-equation system in which the level of defence spending of a country is a function of its adversary's defence spending and other exogenous factors. In the following section, we would like to estimate Eqs. (11) and (12).

EMPIRICAL EVIDENCE

The previous model highlights that the values of disputed territories or resources, along with losses from conflict and the preferences of countries as represented by the parameter β are determinants of the level of military spending. For empirical purposes, we use variables that can serve as proxies for these factors. A more general specification of the utility function given in Eqs. (3) and (4) would yield that we should also account for the size of the economy, which we have denoted by E which for empirical purposes will be accounted for by the real GDP per capita of a country. For example, Seiglie (1992) finds that an increase in the real income of a country results in an increase in their level of real military spending. Therefore, including this variable in a regression on the amount spent on the military we should find that its coefficient is positive in sign.

In addition, estimation of the reaction functions given by Eqs. (11) and (12) indicates that the level of the adversary's military spending is important. Therefore, we will include as an explanatory variable the level of military spending by a country's main adversary. In order to identify a country's main adversary during the time period analysed we proceed as follows. First, we used the 10 Million International Dyadic Events data set established by King and Lowe (2003) that documents all reported international interactions or events between countries from 1990 to 2004. We then calculated a numerical score for all these events using the Goldstein score (see Goldstein, 1992) and for each country we defined its main adversary as the one that it had the highest average conflict score during this time period. For the member countries of NATO, we chose Russia as the main adversary.

There are other variables that have been found to have an effect on the size of the military sector. For example, for countries that are members of a military alliance increases in their allies' expenditures should reduce their 'own' spending if free-riding exists as discussed by Sandler (1993) and Seiglie (1992).[1] Therefore, we include other alliance members' military spending as an explanatory variable in our empirical estimation of the determinants of military spending and expect that the coefficient for this variable is negative. Finally, we will also account for the size of a country by including its population as an explanatory variable since income is being accounted for on a per-capita basis. We would expect that countries with larger populations should spend more on defence. The data for GDP in Purchasing Power Parity terms and population is taken from the World Penn Tables.

As a proxy for the preferences of a country we include several variables that have yet to be fully explored in the literature. The first is the proportion of seats held by women in national parliaments. This data comes from the World Bank's World Governance Indicators. The hypothesis to be tested is whether women have different preferences for military spending than men and this is reflected in the legislative process at the national level when the military budgets are being debated. The next two variables also aim to capture characteristics of nation's economy and politics. The first is the degree of income inequality of a country as measured by the Gini coefficient. Although as mentioned earlier, the effects of real per income have been explored in the literature, the distribution of this income has generally been neglected. The data for the Gini coefficient comes from the World Bank. Finally, we capture an aspect of a country's politics by using a measure of a country's political stability from the World Bank's *World Governance Indicator* (Kaufmann, Kraay, & Mastruzzi, 2010). As mentioned by the authors of this measure, this variable captures 'the perceptions of the likelihood that the government will be destabilised or overthrown by unconstitutional or violent means, including politically motivated violence and terrorism' Finally note that the empirical test is carried out at the country level.

The sample employed is for the 146 countries where data is available during the time period from 1998 to 2007. This period was chosen since it allows for the use of Stockholm International Peace Research Institute (SIPRI) data on military expenditures which has been recently revised. Table 1 presents summary statistics for the variables. The average level of real military spending is $9,926 million for these countries with a large standard deviation of $45,526 million reflecting the large differences in military spending around the world. The average level of real GDP in purchasing power parity from the World Penn Tables is $11,223 million. Furthermore, the average level of an adversary's military expenditures is $64,758 million since for a vast number of countries the adversary is Russia. As in all the military spending data, the standard deviation is quite large. The allies' military expenditures is defined as total NATO spending minus the specific country's spending which measures the degree of spillover from being a member of the NATO alliance. The average spillover is $60,758 million.

Using a fixed-effects model, Table 2 reports the results from regressing a country's military spending in constant dollars on per-capita GDP in constant dollars, their enemy's military expenditures, as well as their ally's, the country's population, as well as a series of variables that capture preferences and domestic factors. These include the proportion of women in the country's Parliament, the country's level of political stability, and finally, the level of income inequality in the country as measured by the Gini coefficient. These estimates are corrected for the presence of

Table 1. Summary Statistics.

Variable	Mean	Standard Deviation
Real military expenditures	9,926	45,526
Population	53,657	181,436
Real GDP per capita	11,223	11,606
Ally's military expenditure	60,758	200,515
Adversary's military expenditure	64,439	81,636
Percent women	14.6	9.75
Political stability	46.4	27.8
Gini coefficient	40.4	9.72

Notes: The data for military expenditures are from SIPRI. *World Military Expenditures and Arms Transfers*. Data are in millions of constant US dollars. GDP and population data is taken from the World Penn Tables. All other data is from the World Bank. The number of observations is 986 for all variables with the exception of the Gini coefficient where because of lack of data the number is 888.

Table 2. Results of Regressing Real Military Expenditures using Pooled Data from 1998 to 2007.

Independent Variables	ln (Expenditure) Fixed-Effects Model	ln (Expenditure) Fixed-Effects Model
GDP	2.83×10^{-5}**	3.42×10^{-5}**
	(8.39)	(8.86)
Ally's expenditure	-1.03×10^{-6}**	-1.24×10^{-6}**
	(3.62)	(4.28)
Adversary's expenditure	1.46×10^{-6}**	1.40×10^{-6}**
	(4.54)	(4.12)
Percent women	0.005**	0.005**
	(2.09)	(2.22)
Political stability	0.003**	0.004**
	(2.75)	(3.05)
Gini coefficient		−0.02**
		(4.62)
Population	4.40×10^{-6}**	4.03×10^{-6}**
	(3.13)	(2.88)
Number of observations	986	888

Notes: The *t*-ratios are shown in parentheses below coefficients.
*Significant at the 10% level.
**Significant at the 5% level.

heteroscedasticity by taking the log of the domestic military spending variable and we estimate the model using fixed effects. Since the data on the Gini coefficient is not available for all countries and time periods we show the results with the exclusion of it as well.

For the first regression, the coefficient for GDP is positive as expected and significant at the 5% level. Evaluating this value at the mean, this implies an income elasticity for military spending of 0.32. That is an increase of 10% in per-capita income results in a 3.2% increase in defence spending. We find evidence of free-riding as the coefficient for ally's military spending is negative and statistically significant, though the effect is small during the post-Cold War period. In terms of elasticity, an increase in other NATO member countries spending by 10% leads to a six-tenth of 1% decline in a member's spending. The coefficient for enemy's military spending is also positive and highly significant. A similar finding holds for the estimated coefficient for population.

Interesting though are the estimates for the percentage of women in a country's parliament. For this variable, increases in women's participation in government leads to increases in military spending. Therefore, it seems that greater women's participation in government does not reduce a country's attitude towards the military as measured by the amounts these governments spend on defence. Similarly, we find that more politically stable countries spend more on defence to protect their 'tranquility' The result can be reflecting the fact that more stable countries have more to lose from international conflict and therefore spend more on defence. We cannot rule out that the causality runs the other way. For example, it is possible that military spending, a portion of which could be used for internal security purposes, leads to more internal order for fear of the consequences of upheavals. Finally, the coefficient for the Gini measure of income inequality in a country is negative and statistically significant at the 5% level. The interpretation of this result is that the greater the amount of income inequality in a country the lower is the amount they spend on the military. Although we have controlled for political stability, this can still be reflecting domestic factors that generate a demand for military spending. For example, countries that have greater income inequality may be concerned with reducing this inequality and may be less inclined to divert government resources away from social programs such as education and towards defence spending.

CONCLUSION

Looking at data from the post-Cold War period, we find that such factors as income, ally's spending on the military, as well as the military spending of an adversary helps to explain a significant portion of the variation in defence spending around the world. Furthermore, we find that other factors such as income inequality, the political stability of a country and the sexual composition of the legislature are also important determinants. For example, we find that greater political stability leads to greater military spending. Another interesting result is that an increase in income inequality actually reduces defence spending. Although our model does not guide us in understanding the mechanism for this empirical finding, we can conjecture that countries with greater income inequality will decide to devote more government resources away from defence and towards social spending, such as education, to reduce this income inequality. We hope to explore this finding further in the future.

NOTE

1. The model presented also yields this result if we modify the contest-success functions given by Eqs. (1) and (2) by introducing a third country allied with either A or D. For example, if we denote the ally by N and it is allied with D, then Eq. (5) becomes $r = \frac{F_D + F_N}{F_D + F_A + F_N}$. The resulting Nash equilibrium level of military spending for the defender declines.

REFERENCES

Dunne, J. P., Nikolaidou, E., & Mylonidis, N. (2003). The demand for military spending in developing countries: A dynamic panel analysis. *Defence and Peace Economics*, *14*(6), 461–474.

Goldstein, J. S. (1992). A conflict-cooperation scale for WEIS events data. *Journal of Conflict Resolution*, *36*, 369–385.

Griffin, L. J., Wallace, M., & Devine, J. (1982). The political economy of military spending: Evidence from the United States. *Cambridge Journal of Economics*, *6*(1), 1–14.

Grossman, H. I., & Kim, M. (1995, December). Swords or plowshares? A theory of the security of claims to property. *Journal of Political Economy*, *1*, 275–288.

Grossman, H. I., & Kim, M. (1996). Predation and production. In M. R. Garfinkel & S. Skaperdas (Eds.), *The political economy of conflict and appropriation* (pp. 73–95). New York, NY: Cambridge University Press.

Gupta, S., de Mello, L., & Sharan, R. (2001). Corruption and military spending. *European Journal of Political Economy*, *17*(4), 749–777.

Kaufmann, D., Kraay, A., & Mastruzzi, M. (2010). *The Worldwide Governance Indicators: Methodology and Analytical Issues.* World Bank Policy Research Working Paper No. 5430. Retrieved from SSRN: http://ssrn.com/abstract=1682130

King, G., & Lowe, W. (2003). An automated information extraction tool for international conflict data with performance as good as human coders: A rare events evaluation design. *International Organisation*, *27*, 617–642.

Maizels, A., & Nissanke, M. K. (1986). The determinants of military expenditures in developing countries. *World Development*, *14*(9), 1125–1140.

Sandler, T. (1993, September). The economic theory of alliances: A survey. *Journal of Conflict Resolution*, *37*(3), 446–483.

Seiglie, C. (1988). International conflict and military expenditures: An externality approach. *The Journal of Conflict Resolution*, *32*(1), 141–161.

Seiglie, C. (1992). Determinants of military expenditures. In W. Isard & C. Anderton (Eds.), *Economics of arms reduction and the peace process.* Amsterdam: North-Holland.

Seiglie, C., & Liu, P. (2002, December). Arms races in the developing world: Policy implications. *Journal of Policy Modeling*, *24*, 693–705.

Smith, R. (1980). The demand for military expenditure. *Economic Journal*, *90*, 811–820.

Smith, R. (1989). Models of military expenditure. *Journal of Applied Econometrics*, *4*, 345–359.

Smith, R. (1990). Defence spending in the United Kingdom. In K. Hartley & T. Sandler (Eds.), *The economics of defence spending: An international survey* (pp. 76–92). London: Routledge.

DEFENCE FIRMS BEYOND NATIONAL BORDERS: INTERNATIONALISATION OR MULTI-DOMESTIC APPROACH?

Renaud Bellais and Susan Jackson

ABSTRACT

Since the end of the cold war, the arms industry has been experiencing a deep structural transformation. The logical consequence of such short-term and structural evolutions is that companies in the arms industry are looking to access markets beyond national borders. Such transformation becomes a policy concern, since the core regulatory framework has been built on a national basis and there are very limited structural and regulatory means for dealing with an international, or even a transnational, industry in such a sensitive activity as weapons development and production. This chapter wonders whether it is possible to qualify such transformation of the arms industry as an internationalisation process and whether it can be compared to the globalisation of commercial activities. It compares what has happened and continues to happen in the arms industry with the globalisation experienced by commercial industries.

The Evolving Boundaries of Defence: An Assessment of Recent Shifts in Defence Activities
Contributions to Conflict Management, Peace Economics and Development, Volume 23, 231–249

ISSN: 1572-8323/doi:10.1108/S1572-832320140000023015

As arms-producing firms have been growing beyond national borders, how can we characterise the restructuring of the arms industry? Indeed, these firms have been for much of their existence centred on a national logic, in which their domestic markets have been predominant and exports have primarily been a means to enlarge production series and improve profitability. However, the situation has been changing since the 1980s, as firms have been bound to adjust to economic and budgetary trends in the post-Cold War era. Arms-producing firms have moved beyond cooperative projects at the international level and have begun a reorganisation beyond national borders, especially large arms-producing companies based in Western Europe.

Such transformation raises questions about the possible emergence of transnational firms and their escape from national regulatory frameworks. At the same time, arms production has remained under strict political surveillance. Mergers and acquisitions are of particular interest in understanding the structural shift. Some researchers do not hesitate to speak about the globalisation of arms production, much in a similar way as that experienced in commercial industries. This assumption begs the question of whether it is possible to make such a comparison.

Nevertheless, the commonly used concept of globalisation can be misleading when it is not actually defined, least not when concerning arms production. What is meant by globalisation in terms of arms-producing firms? Is internationalisation a more accurate portrayal of the post-Cold War structural changes in the arms industry, especially in Western Europe and the United States? Can one identify the features of the globalisation process in the arms industry? This chapter considers the various dimensions of the transformation of this industry since the 1990s. It aims at determining which concepts are the most appropriate to understanding this cross-border transformation. Such selection of concepts can assist in describing the features of the ongoing transformation of arms production. Our chapter aims at identifying the stakes in the debate and focuses on three main players to deepen the analysis: BAE Systems, Thales and EADS (renamed Airbus Group in January 2014).

LEAPFROGGING NATIONAL BORDERS

Facing contracting domestic markets, arms-producing companies have been looking for new markets. Even though exports constitute an alternative, this is quite a limited and competitive market compared to domestic orders in arms-producing countries. While national mergers and acquisitions (M&A) occurred, the biggest potential was provided at the European or international level.

Post-Cold War Trend in the Evolution of the Arms Industry

There are several factors that have led to the transformation of the arms industry since the early 1990s. Most immediately following the Cold War, the industry had been subjected to decreases in military expenditure at the same time the demand for weapons declined. Moreover, technology has become an increasingly important and expensive element of weapons systems. Together, these factors have led to more national and cross-border consolidation, an increase in international collaborative programs and, in some places, a focus on multi-domestic firms.

In the early 1990s, the arms industry experienced several developments that contributed to its initial transformation following the end of the Cold War and the concomitant changes in threat perception. More broadly, there was a global economic recession and a drop in military expenditure leading to a general decline in demand for arms. In addition, client states of the former superpowers rapidly lowered their demand for weapons in part because of the decrease in access to concessional terms for weapons acquisition (Krause, 1994). For some of the larger client state importers during the Cold War, arms trade fell by nearly two-thirds between 1987 and 1991 (Krause, 1994, p. 16).

Another pivotal development that contributed to the transformation of the arms industry is the rapid changes in technology that prompted technology to become a spin-in element rather than a spin-off. That is, technological developments in the civilian sector are now integrated into weapons programs rather than vice versa as had been during and immediately following the Cold War. Further, due to its increasing importance in weapons systems, technology began to drive systems costs. The rate of growth in cost has been faster for weapons that incorporate or rely on new technologies. These new technologies are not only more expensive to manufacture but also require more research and development, more highly trained operators and more expensive spare parts, each factor adding to the overall higher cost of the end product. Even with the wider use of off-the-shelf commercial components, technology pushes these price increases in weapons systems (Neuman, 2010). This push from technology has only continued to rise in more recent years, which contributes to the difficulty for smaller national markets and industries to maintain or develop competitive arms industries. The expense of technology is a key factor driving transformation in the arms industry making it nearly impossible for arms producers, except in the United States, to be able to afford their own production of a comprehensive range of weapons systems because of the cost of manufacturing high-technology systems to replace their conventional forces (Dunne & Surrey, 2006). Technology also creates an entrance barrier for arms producers that do not already have large scale and technological maturity (Matthews, 2004).

Both trends resulted in a 'scissor effect' where arms-producing firms were developing increasingly costly systems but in a context of decreasing national budgets. This situation was bound to result in a contradiction between the size of the firm and their then available market. In this way, economic trends have been challenging the sustainability of firms on their domestic national market. Arms industry consolidation appeared as the best means available to adjust the number of players towards what was typically a monopoly in terms of system type. Nevertheless, in some cases, even a national monopoly was not sustainable. At that point, the consolidation process entered into an international phase at the European level or towards a more transatlantic orientation for Western European firms.

Over the last two decades, the majority of firms in the United States acquired by European firms have been mixed commercial and arms producers though there were a few large acquisitions of companies categorised as primarily arms producers. On the whole, most of the European acquisitions in the United States were sub-system firms and suppliers in the second and third tiers rather than prime contractors. American company acquisitions in Europe have been mostly in the United Kingdom. As with European acquisitions in the United States, there were few American acquisitions of prime contractors in Europe and most of these acquisitions involved dual-use businesses and companies with a more commercial focus (Bialos, Fisher, & Koehl, 2009). By 2009, arms-producing companies were continuing to diversify into the security industry more broadly including cybersecurity and intelligence acquisitions.[1] Many of these acquisitions were of American companies, acquisitions that then provide potential access to the American arms market. In addition to outright acquisitions, many firms have sought shares in foreign companies, which in turn have led to changes in ownership structures (Dunne & Surrey, 2006).

By the mid- to late 1990s, mergers and acquisitions (M&A) had increased and consolidation was encouraged especially in the United States.[2] In the late 1990s, the Clinton Administration in the United States seemed amenable to transatlantic M&A, at least in rhetoric if not actual practice. However, because of the Bush Administration's priorities, the events of 11 September 2001 and the Europe and United States rift over both the Iraq war and the Dubai port controversy, by the early years of the 21st century, the United States had become more restrictive in rules involving assets considered having to do with national security. This more restrictive approach began to loosen midway in the Bush years to what has emerged today (Bialos et al., 2009). In addition, more recent merger and acquisition activity has been at the small company/supplier level and is reflective of the need to acquire access to technology more than the desire for growth as it was previously (Dunne & Surrey, 2006).

International collaborative programs between firms were set out to deal with smaller national markets and the increasing expense of technology. These programs take many forms of which a variety can be utilised in any given transaction, including joint ventures, teaming arrangements, co-prime arrangements and greenfield manufacturing operations. Transatlantic collaborative programs largely involve subsystems work (Bialos et al., 2009). In the United States, arms producers often focused on competences, selling off what did not support these capabilities and joining in collaborative efforts to lower R&D costs and to pool together orders to lengthen production runs (Dunne & Surrey, 2006). Similarly, these programs also often have involved European firms occupying niche markets and supplying subsystems to US-industry led projects in order to maintain or develop a core competence for domestic industry (Neuman, 2010) though programs also can be between European firms (Krause, 1994). Whatever the types of collaboration, these niche markets are targeted to assist firms in achieving economies of scale (Krause, 1994).

One strategy employed by especially European arms producers in order to counter smaller domestic markets and the need to diversify their customer base is the multi-domestic firm approach, that is, having multiple home markets (Bialos et al., 2009). This type of approach contributes to an increase in the international component of these firms' turnover. For example, UK's BAE Systems has invested in home markets outside of the United Kingdom, of which it currently has five (Australia, India, Saudi Arabia, the United Kingdom and the United States). Its US operating groups – the company's two largest operating groups – alone generate 45 per cent of

BAE's total revenues (BAE Systems, 2012, p. 125). In 2011, BAE Systems generated £3.75 billion in revenues in the United Kingdom where as it generated a total of two-thirds in its other home markets combined, £7.29 billion of that in the United States. Typically, home markets are run on a national rather than global basis (Bialos et al., 2009) with decisions made accordingly.

When we look at BAE Systems, Thales or EADS, such evolution has transformed the relationship of the firm to the national states. Until the 1980s, key founding firms of these groups – British Aerospace, Thomson-CSF and Aerospatiale, respectively – were state-controlled firms which were strongly dependent on their domestic market and the support of their home country for export markets. There has been a revolution of sorts in the structure of the arms industry since the 1980s and the withdrawal of state capital from many arms producers. British Aerospace (now BAE Systems) was the first large industry to be auctioned off in 1981. Rolls-Royce, nationalised in 1971, was privatised again in 1987. Further UK privatisations involved the Royal Ordnance Factories, Short Bros (Bombardier) and DERA (QinetiQ). A similar trend occurred in France, with a retreat of the state's ownership through waves of mergers and consolidation: Thomson-CSF (now Thales) in 1997; Aerospatiale in 1998 (merger with Matra) and 2000 (creation of EADS); Snecma in 2004 (merger with Sagem to create Safran); and DCNS in 2005 (25 per cent stake sold to Thales).

Strategies of diversification into commercial activities and internationalisation of defence assets have undermined the traditional state–firm relationship. Indeed, a reduced weight of domestic arms orders makes these firms less dependent on the defence policy of the state where the firm is headquartered and diversified arms markets can reduce the market power of a given state when it comes to arms procurement. Therefore there is little to compel larger, more diversified and more international groups to follow the guidelines fixed by, here, the British or French state rather than create a more autonomous strategy.

One concern may be that arms-producing companies are then without governmental oversight and restrictions and therefore uncontrollable. Nevertheless there is a misleading use of the concept of globalisation, since researchers do not have a consistent (or even clear) definition of this term: it is possible to mix the transformation of markets, firms' strategies to access to markets and the industrial and technological base. It is therefore necessary to examine which concepts can be used to analyse the ongoing transformation of the arms production before assuming the arms industry has been globalised.

Internationalisation or Globalisation?

It is difficult to understand transformation in the arms industry in terms of economic globalisation. Indeed, it is not easy to qualify transformation in the case of commercial industries. There are several political economic concepts utilised to attempt to capture what has been going on in the economy since the 1980s, and it is common to mix different trends under the generic concept of 'globalisation'. Indeed most authors tend to use such term as a synonym for multi-nationalisation or inter-nationalisation, and at times are simultaneously referring to both demand and supply sides.

This confusion creates a difficulty when trying to understand the underlying dynamics in the transformation of the arms industry and the possible consequences of this transformation. In this chapter, we focus on the supply side – the arms industry – and the capital and production dimensions of firms' international strategies. If the globalisation can be associated to the opening of markets and their increasing integration, the evolution of the industry concerns both access to markets and the location of production. Therefore, when using the economic literature, it is possible to distinguish between internationalisation and globalisation.

We can define the *internationalisation* of the industry as the ability of firms to access markets beyond their own domestic market through either exports or/and investment. Here, exports represent the initial, minimal step in the growth strategy, which does not require that a given firm deploy its production beyond national borders. In the arms industry there is nothing really new in this respect. At the beginning of the 20th century for instance, Vickers was exporting a large share of its production abroad through the effective action of Basil Zaharoff (Bellais, 2000). Arms trade is an even longer-term trend since, as Krause (1992) underlines, exports have been almost consubstantial to the arms industry since the Renaissance.

In fact, firms can also need to acquire productive capacities in other countries either to complement their existing capabilities or get closer to their foreign customers. Their production becomes international as well as their turnover, but this does not mean that they have a global approach. Being international here only means that a given firm has productive capacities in more than one country. There is a key performance indicator to assessing internationalisation in the arms industry – turnover – which can be broken into two elements according to demand and supply (Dorrenbacher, 2000). Demand-based turnover is measured by the amount of turnover generated by activities in foreign countries. Supply-based turnover has to do with how much a company's foreign subsidiaries generate turnover. These types of turnover measure what kind of impact a company's foreign presence has on overall company performance, which in turn indicates the degree to which a company has experienced internationalisation.[3]

In addition, many firms see greenfield manufacturing facilities as a means for increasing market access in foreign markets. This strategy is especially appealing to European arms producers as a means for entering the US market (Bialos et al., 2009). These facilities, as the argument goes, are attractive because home governments see them as a means for maintaining domestic employment while allowing for more open competition involving foreign firms.

At the other end of the integration spectrum, *globalisation* constitutes a path-breaking change since firms no longer adjust their strategies in accordance mainly to political borders and they reorganise their productive capabilities on a global perspective. This is what Friedman (2005) means when he states that since the end of the 20th century 'the world is flat'. Even though political borders still exist, they do not represent a key parameter when firms determine strategic decisions related to productive capacities. In a globalised approach the world represents a unique market and a unified market in which firms manage their means and targets (though in most industries, globalisation in its purest sense has not occurred).

In fact between both extremes, there are several possible configurations in terms of international strategy at the capital and production levels. What then are the most relevant concepts when discussing arms producers? As underlined by this brief overview of the distinction between internationalisation and globalisation, it is crucial to have a clear lexicon to avoid misunderstanding in the analysis presented here.

One can use the taxonomy proposed by Hagedoorn and Schakenraad (1994) to distinguish the various possible configurations of the arms industry. Indeed, there are numerous academic frameworks to apprehending firms' strategies. In particular, Hagedoorn and Schakenraad's matrix is both simple and comprehensive in order to handle the transformation of the arms industry. This matrix differentiates several modes of company organisation with respect to production and markets (Table 1).

National or *domestic* firms produce in one country and serve mainly that particular domestic market. Such category corresponds to several arms firms, even today, and can represent the starting point of our questioning in this chapter. The internationalisation of such firms can be understood as an expected extension of the domestic market once it has been saturated or is not large enough to allow firms to grow or just survive. However,

Table 1. A Differentiation of (Inter-)Nationally Operating Firms.[5]

		Markets		
		National	International regional	Worldwide
Production	National	Domestic	Semi-international	International
	International regional	–	Multi-domestic	Multinational
	Worldwide	–	Transnational	Global

Source: Hagedoorn and Schakenraad (1994), p. 190.

semi-internationalisation or full *internationalisation* does not structurally change these firms. Even though they enlarge their market and turnover, they do not need to change their productive organisation. Here the internationalisation process only raises issues in terms of export control.

What has potentially more impact on the structure of firms is when firms expand their productive capacities beyond their national borders through acquisitions or greenfield investment. Here Hagedoorn and Schakenraad propose four categories, though these can be gathered into two core trends linked to the productive organisation of these companies.

Multi-domestic and *multinational* firms correspond to an internationalisation of productive resources but most of the time to a limited extent. Indeed, even though some industrial consolidation might exist in these configurations, one may perceive, with the prefix 'multi', that there is a consolidation of capitalistic entities rather than a rationalisation of industrial capabilities or/and a duplication of a given capability into another domestic economy. This type of consolidation is especially the case when talking about arms producers, which are most often multi-domestic or multinational in the sense that they own capabilities in different countries but without going the step further, that is, the consolidation of industrial and technological capabilities across borders.

Transnational and *global* firms represent fully achieved internationalisation since this process includes not only sales but also productive and technological capacities. In such a configuration a firm aims at organising its productive resources worldwide and is able to set up centres of excellence at the global level: 'Global companies as the true opposites of domestic companies [...] operate on a worldwide scale with respect to both their markets and their production' (Hagedoorn & Schakenraad, 1994, p. 167).

Such a configuration is interesting, for even the nature of trade changes. For instance, Grossman and Rossi-Hansberg (2008) demonstrate that in a truly globalised economy, trade is not only constituted of goods but also of tasks, since the organisation of production is optimised at the global level to match technology constraints and market sizes. Baldwin (2006) even remarks that globalisation corresponds to the second unbundling of the economy, which represents the triumph of Adam Smith (social division of labour) over David Ricardo (comparative advantages).

Then the question becomes: Are we able to assess the internationalisation or globalisation of the arms industry? Can we find the appropriate information to estimate the depth (or lack thereof) of its globalisation?

CONTRADICTORY DYNAMICS OF INTERNATIONALISATION

Even though M&A could favour the creation of transnational groups, there are strong limits to the integration. The arms industry has been experiencing a specific form of internationalisation, which results from the political dimensions related to its products and services. As demonstrated in the failure of the EADS/BAE Systems merger in 2012, this industry remains highly regulated and any change beyond national borders is very unlikely to succeed without a strong and clear political blessing (Bellais & Droff, 2013).

From National to Multi-Domestic Firms

There are several specific dimensions through which to examine and analyse internationalisation and globalisation in terms of the arms industry. These include capital, sales and production capabilities (including R&D). Indeed, the internationalisation process is neither uni-dimensional nor homogenous; one can expect that it can progress at various rhythms and in different forms in accordance with which dimension is analysed.

First, the internationalisation of firms can go through mergers and acquisitions (M&A) in a purely financial dimension, that is, via ownership. Capital seems to be the most open in terms of cross-border transactions though this openness varies by country and is often dependent on the location of the investor. That is, there are varying levels of access to foreign ownership in different countries. One motivation for cross-border M&A is to gain market access. This capital consolidation of arms production across borders does not necessarily mean, though, that a company is bound to consolidate other dimensions of its activities. Such M&A can be limited to an investment strategy in a portfolio approach without any industrial dimension – similar to what a financial investor could choose.

Second, internationalisation can imply the interpenetration of markets. Here this process aims at enlarging the sales base of a given firm. When its domestic market is not sufficient to support its growth or even to help it remain sustainable, internationalisation can open new markets and develop a strategy of economies of scale and/or scope. Production has more limited flexibility in terms of location, though growing export shares as well as greenfield investments both contribute to a broadening of production locations.

Once a firm is internationalised in this dimension, it becomes more difficult to distinguish what domestic sales and exports are, since the firm can have several domestic markets and often produces in each as well as exchanges goods between them.

Third, firms can deepen their internationalisation by reorganising their production assets (and even R&D capabilities) across borders, especially in order to suppress excess capacities and improve the effectiveness of production in pure industrial terms by developing economies of scale (reconciliation of production volumes with the optimal production by site) and scope (creation of centres of excellence). One may consider this step as the culminating stage in internationalisation, which characterises globalisation in its specific nature.

Once these levels defined, it appears that arms production is far from the depth of the internationalisation process experienced by (at least some) commercial industries and it is even farther from globalisation. There are only a few examples of production consolidation in the arms industry that have resulted from M&A, for example, Eurocopter (renamed Airbus Helicopters in 2014), AgustaWestland, MBDA, Astrium (now part of Airbus Defence and Space) and Thales Alenia Space. However, even in these cases, the production integration has not been fully achieved and many activities in production and even more in R&D remain structured in their initial national configuration.

What would have been a logical trend for the supply side has been strongly limited by the demand side, that is, those states where activities are located. Indeed states have two tools to control the production restructuring:

- They remain the largest customer and can 'pilot' the industrial setting when deciding where production is located (for instance, applying rules

such as the national security exception which encompasses the 'security of supply' or 'strategic autonomy'). This piloting helps explain why, for instance, satellite producers maintain integration capabilities in each country as well as why helicopter firms have so many final assembly lines.
- Any industrial restructuring across borders supposes that firms are able to exchange components between countries in a task-trading dynamics (Baldwin, 2006; Grossman & Rossi-Hansberg, 2008). However, this exchange requires a permissive pre-condition in arms production for such dynamics: there is no possible trade without export-control authorisations.

It is therefore understandable why cross-border consolidation in arms production has led to multi-domestic groups, lest national states themselves had pushed for a deeper consolidation at the production level as in the case of space and missiles fields in Europe (as it was the only condition for preserving local production capabilities). When speaking with insiders, it becomes clear that EADS or Thales are mainly multi-domestic groups and it is more evident for BAE Systems when one analyses the organisation of its productions between both sides of the Atlantic. Table 2.

Production capabilities are still very much based on national schemes and lack overall integration. Together and separately these dimensions allow researchers to assess how much the arms industry has internationalised and whether that is leading to globalisation as can be seen more so in commercial industries. At the same time, even with mergers and acquisitions, industrial capability remains largely non-integrated.

An Internationalisation without Industrial Consolidation?

Even though internationalisation in the arms industry has been ongoing since the 1990s, it is difficult to demonstrate that such a process is leading to a globalisation as is the case in many commercial activities. Indeed, political regulatory frameworks remain very effective at the level of productive and intellectual resources and one can perceive a heterogeneous evolution: firms need to access non-domestic markets to sustain their turnover and promote their growth, but they have a limited margin of manoeuvre to deepen their market strategy at the production level to consolidate their productive capabilities.

In fact one can note that control mechanisms have been reinforced since arms producers started internationalising. As long as firms limited their

Table 2. Depth of Internationalisation.

Firm	Ownership	Sub-System Production	Integration/Final Assembly Line	R&D
Astrium (EADS)	Fully consolidated	Partial specialisation	Nationally based	Nationally based
BAE Systems	Fully consolidated	Chinese wall between the UK and the US	Nationally based	Nationally based
Eurocopter (EADS)	Fully consolidated	Partial specialisation	Nationally based	Nationally based, Limited specialisation
MBDA (EADS, BAE Systems, Finmeccanica)	Fully consolidated	Partial specialisation (ex. Scalp Stormshadow)	Nationally based	Nationally based, limited specialisation
Thales Alenia Space (Thales, Finmeccanica)	Partially consolidated	Partial specialisation	Nationally based	Nationally based
Thales radars	Fully consolidated	Nationally based	Nationally based	Nationally based

Sources: Insiders' Interviews.

international activities to exports and had mainly national stockholders, regulation focused on exports and technology transfers. Since the early 1990s, states have implemented new tools to control firms' shareholding, validate foreign acquisitions, supervise decision in terms of industrial restructuring and so on (Bellais, 2010).[4] The new regulatory framework tries to adapt to the structural transformation of the arms industry. Thus firms are in a paradoxical situation where their adjustment over national borders is step-by-step controlled in its industrial dimensions by national decisions.

In a sense, this situation can be compared to what some consider as the first globalisation of the world economy (i.e. before 1913) because that period was characterised by an internationalisation through the exposition of flows of goods, not mainly through productive investment or assets. As discussed above, this type of internationalisation was especially the case for the arms industry. Leading firms such as Vickers in the United Kingdom, Krupp in Germany or Schneider in France were internationalising their sales, but this trend was mainly supported by the opening of foreign markets (Russia, Japan, Balkans countries, etc.) and with a limited number of foreign subsidiaries.

Nevertheless one should not overestimate the depth of globalisation of productive structures, even in commercial activities. Many economic studies on the globalisation process note that there are very few truly globalised firms (Mayer & Ottaviano, 2007). In a general perspective, Perraton (2001, p. 672) notes that 'national economic systems remain central for generating specific patterns of production and economic performance'.

This is especially true concerning arms production. Even though firms have been engaged into a capitalistic consolidation, this latter is submitted to ex ante approvals from states and, once the consolidation is achieved, states remain the key players to favour or prevent firms from reorganising their productive means across borders. While firms mainly structure markets and productive organisations in commercial industries, in the arms industry the demand side (states) still pilots the organisation of productive structures, that is, the supply.

If one expects to qualify today's international configuration of arms firms, it could be useful to refer to Pankaj Ghemawat's concept of semi-globalisation, even though he only deals with commercial markets. Indeed arms firms can be neither fully globalised nor only localised. Ghemawat defines semi-globalisation as 'a situation in which neither the bridges nor the barriers between countries can be neglected' (Ghemawat, 2003, p. 5). His description characterises an intermediate level of cross-border market integration.

Even in commercial markets globalisation is difficult to reach since markets are heterogeneous and firms need to adapt both commercial strategy and production to local features. On one side, firms try to optimise their strategies beyond national borders, especially to enlarge their markets and optimise production or sourcing through economies of scale. On the other, notes Ghemawat (2007), they have to take care of the specific features of national markets and consequently adjust both their commercial and industrial strategies in each country.

It is to a certain extent the case in arms production. States tend to require some local content in the weapons and equipment they buy. The world is not flat, that is, it is not homogenous when it comes to the arms industry; therefore, national boundaries remain important because of national regulation and the related specificities of national markets. For this reason, most arms firms can be considered as multi-domestic rather than globalised. They have acquired other companies at the financial level but such M&A have not led to an industrial consolidation. These M&A have resulted in groups which operate in different countries but whose operations are not necessarily linked across borders.

Even though the internationalisation of arms firms can go through exports or acquisition of foreign producers, this strategy does not solve the core problem of many firms. In fact, beyond the growth strategy, internationalisation constitutes one answer to productive constraints: given today's technology, most domestic markets are too small to allow even a domestic monopolist to survive by relying only on its own domestic market. Therefore, industrial consolidation, beyond a capital one, is an underlying factor of a sustainable defence industrial base, as basic industrial economics would argue.

However, as we have underlined, in almost all the cases, political and regulatory barriers prevent arms firms from reorganising their productive assets over national borders. One almost unique exception is MBDA, which has started to rationalise its industrial capabilities at the European level a decade after it was created. Nevertheless, here it is an ultimate survival decision from the demand side, which would have lost any European industrial missile capabilities otherwise.

It is then like trying to square the circle. On one side, domestic markets are no longer big enough to support one firm in a given market segment; on the other, firms' strategies are caught by national regulatory frameworks which prevent them, at least in medium and large arms-producing countries, from deepening the consolidation beyond the capital level. Is there another alternative? One may consider the role played by cooperative programmes.

Cooperative programmes can indeed be considered as a substitute to globalisation. In a sense, if such programmes are set up to optimise industrial organisation, the sharing of work packages is likely to promote the emergence of centres of excellence rather than duplicate productive capacities. Here, cooperation could become a means to achieve task sharing at regional/international level between firms rather than within a given group. Nevertheless, empirical examples demonstrate that concrete experiences are far from promoting such industrial rationale (Hartley, 2008).

CONCLUSION

The French historian Fernand Braudel (1985, p. 103) once stated that a national economy is a political space transformed by the state. Even though arms firms' strategies are less and less connected to states' targets, these latter keep a strong grip over the fate of their arms industry. While firms can be allowed to create capital links across borders, this linkage does not necessarily lead to a consolidation of industrial capabilities contrary to what is often the case in commercial markets.

In a sense arms firms are today in a stalemate, stuck between domestic markets which are most of time too small for a sustainable industry and states which block or prevent the core transformation required at the production level. As underlined by the economic literature, here cooperative programmes are a second-best solution, but it is possible their disadvantages are so large that cooperation is not an alternative to actual industrial consolidation.

One may then wonder whether it is possible to avoid out-of-control industrial consolidation while altering the regulatory framework to promote industrial consolidation in the arms industry. Indeed, arms production does not seem to escape from the basic laws of industrial economics which would argue the best solution is the reconciliation between the size of the market and the technological and industrial considerations. In a sense, the supply side leads such a process in commercial markets, though this trend is prevented by national regulatory mechanisms in arms production.

Is it possible for states to reach such reconciliation? Even if the United States has been able to avoid this question, it is no longer the case for European countries. However, attempts to create a European-wide arms market have failed up to now, for instance with the WEAG, LoI (Letter of

Intent) or OCCAr. Nevertheless, the European Parliament adopted in 2009–2010 a Defence Package, which consists of a possible new regulatory framework at the European level. It will be interesting to see whether such an emerging regulatory framework will lead to a controlled restructuring of the industrial capacities of the arms industry in the years to come.

NOTES

1. The term security industry 'lacks a well-formed and universally accepted definition [but in] the broadest sense of the term encompasses products and services as diverse as alarm systems, electronic access control and biometrics, and security and surveillance consulting' (Jackson, 2010, p. 264).
2. The majority of arms company acquisitions that can be tracked for analysis are national acquisitions in the United States and Western Europe or transatlantic acquisitions between the two. Foreign acquisitions in the United States have been predominantly made by UK arms producers.
3. To measure degree of transnationalisation of commercial firms, UNCTAD utilises transnationality index comprising an average of three ratios: foreign sales to total sales, foreign assets to total assets, and foreign employment to total employment. UNCTAD also has an Internationalisation Index calculated from the number of foreign affiliates divided by the number of all affiliates (UNCTAD, 2008).
4. To varying degrees, countries have laws pertaining to foreign direct investment not least when it comes to investment in arms production or other areas deemed a 'national security' interest. The United Kingdom has the Enterprise Act with several sections devoted to areas of public interest and one in particular that specifies a national security exception and the government's right to intervene in such mergers (Enterprise Act 2002, http://www.opsi.gov.uk/acts/acts2002/ukpga_20020040_en_1). Germany's Foreign Trade and Payments Act and related amendments place a 25 per cent threshold on foreign investments in Germany's arms producers (Thirteenth Act amending the foreign trade and payments acts and the foreign trade and payments regulations of 18th April 2009, www.bmwi.de/.../englischer-gesetzestext-eines-dreizehnten-gesetzes-zur-aenderung-aussenwirtschaft,property=pdf,bereich=bmwi, sprache=...). Similarly, France's Decret No. 2003-196 du 7 mars 2003 regulates these types of foreign investment (Decret No. 2003-196 du 7 mars 2003 réglementant les relations financières avec l'étranger, http://admi.net/jo/20030309/ECOT 0237030D.html).
5. One may wonder why there is no concept for a firm with a purely national market but with regional or global production. The reason is that here we are considering the transformation of the arms industry at the firm level in terms of strategy and internal organisation. The internationalisation of the supply chain represents an interesting issue, but is not considered here to avoid mixing parallel evolutions of firms' strategies and the transformation of value chains in arms production. We therefore do not consider the internationalisation of the external sourcing.

REFERENCES

BAE Systems. (2012). *Annual report 2011: Leveraging global capability*. London: BAE Systems.

Baldwin, R. (2006). *Globalisation: The great unbundling(s)*. Geneva: Graduate Institute of International Studies.

Bellais, R. (2000). *Production d'Armes et Puissance des Nations*. Paris: L'Harmattan.

Bellais, R. (2010). Une industrie de défense en transition, implications pour la sécurité internationale. In J. Fontanel (Ed.), *Économie politique de la sécurité internationale* (pp. 217–236) Paris: L'Harmattan.

Bellais, R., & Droff, J. (2013). La nécessaire réorganisation de l'industrie de défense en Europe. *Revue Défense Nationale*, 757(February), 47–52.

Bialos, J., Fisher, C., & Koehl, S. (2009). *Fortresses and icebergs: The evolution of the transatlantic defence market and the implications for US national security policy*. Washington, DC: Center for Transatlantic Relations.

Braudel, F. (1985). *La dynamique du capitalisme*. Paris: Flammarion.

Dorrenbacher, C. (2000). *Measuring corporate internationalisation: A review of measurement concepts and their use*. Discussion Paper FS I 00-101. Wissenschaftszentrum Berlin fur Sozialforschung, Berlin.

Dunne, J. P., & Surrey, E. (2006). *Arms production*, SIPRI yearbook 2006 (pp. 387–418). Oxford: Oxford University Press.

Friedman, T. L. (2005). *The world is flat, the globalized world in the twenty-first century*. New York, NY: Farrar, Straus and Giroux.

Ghemawat, P. (2003, June). Semiglobalization and competitive strategy. Working Paper. Harvard Business School, Boston, MA.

Ghemawat, P. (2007). *Redefining global strategy, crossing borders in a world where differences still matter*. Boston, MA: Harvard Business School Press.

Grossman, G., & Rossi-Hansberg, E. (2008). Trading tasks: A simple theory of offshoring. *American Economic Review*, *98*(5) 1978–1997.

Hagedoorn, J., & Schakenraad, J. (1994). The internationalization of the economy: Global strategies and strategic technology alliances. In U. Muldur, R. Petrella, & M. G. Colombo (Eds.), *The European community and the globalization of technology and the economy* (pp. 159–192). Luxemburg: Commission of the European Communities.

Hartley, K. (2008). *The industrial and economic benefits of Eurofighter Typhoon*. York: Centre for defence Economics, University of York.

Jackson, S. T. (2010). *Arms production*, SIPRI yearbook 2010 (pp. 251–283). Oxford: Oxford University Press.

Krause, K. (1992). *Arms and the state: Patterns of military production and trade*. Cambridge: Cambridge University Press.

Krause, K. (1994). *The maturing conventional arms transfer and production system: Implications for proliferation control*. Research report for the Non-proliferation, Arms Control and Disarmament Division of the Department of Foreign Affairs and International Trade, York University and Monterey Institute of International Studies.

Matthews, R. (2004). Defence offsets: Policy versus pragmatism. In J. Brauer, & J. P. Dunne (Eds.), *Arms trade and economic development: Theory, policy, and cases in arms trade offsets* (pp. 89–102). London: Routledge.

Mayer, T., & Ottaviano, G. (2007). *The happy few: The internationalisation of European firms*. Bruegel, Brussels: Bruegel Blueprint 3.

Neuman, S. (2010). Power, influence, and hierarchy: Defence industries in a unipolar world. *Defence and Peace Economics*, *21*, 105–134.

Perraton, J. (2001). The global economy – Myths and realities. *Cambridge Journal of Economics*, *25*, 669–684.

UNCTAD. (2008). Largest transnational corporations pursued further expansion abroad, report says, Press release, UNCTAD, Geneva. Retrieved from http://www.unctad.org/Templates/Webflyer.asp?docID=10509&intItemID=4697&lang=1

DEFENCE OFFSETS: REGULATION AND IMPACT ON THE INTEGRATION OF THE EUROPEAN DEFENCE EQUIPMENT MARKET☆

Gueorgui Ianakiev

ABSTRACT

The use of offsets is one of the main characteristics of international defence trade. The rising costs of defence equipment and the significant contraction of defence spending have resulted in an environment that favoured the use of offset policies, the latter becoming increasingly demanding in both quantitative and qualitative terms. The chapter analyses the role of offsets on the process of integration of defence equipment markets, with a specific focus on the EU. Particular attention is given to the offset-relevant regulation and practice and to their recent evolution in the EU following the adoption of European Directive on

☆The views expressed in this chapter are those of the author alone and not necessarily of the European Commission.

The Evolving Boundaries of Defence: An Assessment of Recent Shifts in Defence Activities
Contributions to Conflict Management, Peace Economics and Development, Volume 23, 251–270

ISSN: 1572-8323/doi:10.1108/S1572-832320140000023016

defence and security procurement (81/81/EC). Offsets play a dual role with regard to the integration of defence industries: on one hand they can be trade-distorting and contribute to the survival of inefficient suppliers in arms importing countries; on the other hand, they can contribute in overcoming barriers that may otherwise prevent some potentially efficient suppliers from accessing the supply chains of the big system integrators. The chapter draws the attention on the need to complement the regulatory evolution by further initiatives aiming at improving the access of non-incumbent suppliers to the supply chains of the large defence system integrators.

The use of offset policies has become, over the past decades, one of the most prominent features of international defence trade (For an overview see Martin, 1996; Brauer & Dunne, 2004). The rising unit production costs of defence equipment (Kirkpatrick, 1995, 2004; Pugh, 1986, 1993) and the significant contraction of defence spending that followed the end of the Cold War has resulted in the combination of a higher level of competition for exports contracts between large arms producing countries on one hand, and an increased emphasis on maintaining or building some specific industrial capacities in arms importing countries on the other hand. This environment favoured the recourse to offset policies that progressively became increasingly demanding in both quantitative and qualitative terms.

Offsets can be defined as provisions to an import agreement, between an exporting foreign company, or possibly a government acting as intermediary, and an importing public entity, that oblige the exporter to undertake activities in order to satisfy a second objective of the importing entity, distinct from the acquisition of the goods and/or services that form the core transaction. The incentive for the exporter results from the conditioning of the core transaction to the acceptance of the offset obligation. Offsets must also possess the property of additionality, that is 'the extent to which, as a consequence of the offset obligation, the prime contractor buys from his customer quantities of goods and services, over and above what he would have bought in the absence of the offset agreement'.[1]

The presence of offset requirements is one of the main characteristic features of international arms trade. It also is one of the most controversial ones. Offsets have progressively gained such importance that the debate related to them started obscuring the attention given to the core contracts to which they are attached. A widespread negative preconception appears to be present in the literature leading to the understanding that offsets are distortion-creating practices that are mainly aimed at supporting inefficient local suppliers in importing countries.

As such they could be regarded as a factor of resistance against the integration of the Defence Technological and Industrial Base (DTIB). The issue is of particular importance in Europe, where longstanding national defence industrial policies have resulted in a fragmented market, costly duplications and comparatively small production series with the resulting consequences in terms of unit costs rise. It is important to note the high importance of intra-European arms sales. According to SIPRI data, over the period 2004–2008 for instance, the EU represented by far the largest destination of exports of major conventional weapons by EU exporters with 33%, while Asia, coming second, represented 22% and the Middle East 19%.[2]

Evidence is however also present that under some conditions the use of offsets can lead to the internationalisation of the large system integrator's supply chains through the discovery of new efficient suppliers in arms importing countries that can replace less efficient incumbent suppliers. This duality of the effects of offsets on the integration of the DTIB poses not only an important analytical challenge, but also has very important consequences with regard to the definition of an adequate regulatory framework regarding offsets in Europe.

In section 'Offsets in International Defence Trade' we will characterise the importance of offsets in international defence trade, with specific emphasis on the situation in Europe. Section 'Offset-Relevant Regulation at the International Level: The GATT/WTO' is dedicated to the description of the relevant regulation framework that has existed at the international and at the European level. This leads in section 'Defence Offsets in the European Union' to analyse the use of offset requirements in Europe. Section 'The Challenge of Tackling Defence Offsets' will introduce some recent changes that affect the regulation framework, especially the introduction of 81/81/EC,[3] and section 'The Introduction of Directive 2009/81/EC and its Consequences' will approach the consequences of the changes in regulation on the organisation and on the functioning of the European defence industry and market. Section 'Offsets: A Driver of Integration or Disintegration of the DTIB?' then explores the consequences of the regulatory changes with regard to the objective of integration of DITBs in Europe by taking into account the dual influence of offsets and their potential to be a vector of both integration and disintegration.

OFFSETS IN INTERNATIONAL DEFENCE TRADE: IMPORTANCE AND REGULATION FRAMEWORK

Offsets generate debates in both importing and exporting countries. While importing countries expect to maximise potential benefits for their own economy, these obligations may imply a net cost in export reduction and support to the DTIB of exporting countries.

Offsets in International Defence Trade

Offsets become popular in the 1970s, following, in particular, several highly symbolic agreements such as the joint purchase of F-16 by Belgium, Denmark, the Netherlands and Norway concluded in 1975, or the sale of F-5 combat aircraft to Switzerland also negotiated during the same period. In these transactions, the offset agreements played an important role in both securing the deal for the successful contender and shaping the perception that important economic benefits could be obtained by arms importing countries through compensation requirements. The F-16 deal, often referred to as 'the deal of the century', for instance provided the defence industries of the four European buyers with a participation in the production of every F-16 sold in the world, including those acquired by the US Air Force. It was also instrumental in the huge commercial success of the F-16 combat aircraft.

The general term offset of course covers a wide variety of transaction types. It is possible to establish different typologies of offsets based on their link with the core transaction or on the nature of the compensatory transactions. First, it is possible to distinguish direct from indirect offsets, the former covering activities directly linked to the goods and services that form the object of the main contract. Another distinction can be made between defence-related and civil offsets. Finally, a very large array of activities can be used in order to comply with offset requirements. The list of the different categories of offsets used by the United States (US) public institutions for instance includes the following: technology transfer, Subcontracting, co-production, credit assistance, training, licensed production, investment, purchases and other.

From the 1970s on, offset demands have been systematically increasing, both in value and qualitatively in terms of industrial and technological intensity. The objectives of offset policies vary widely from one country to another. Some countries aim at sustaining a viable defence sector able to sustain the national defence and to limit external dependence. They will normally concentrate on defence offsets, and will put a comparatively high emphasis on direct ones. Other countries may show a higher interest in indirect offsets because of the highest flexibility that they allow in trying to induce the establishment of industrial relation between national enterprises and the foreign system integrators. Finally, some countries that do not possess a sizeable defence industry and are not interested in acquiring capacities in this field may concentrate on civil offsets.

Progressively, an increasing number of countries were introducing official or unofficial offsets policies. The *Offset Guidelines Quarterly Bulletin* published by Countertrade & Offsets (CTO) currently provides information on the offsets policies of 75 countries worldwide. The evolution of the average offset percentage, i.e. the ratio of the value of offset agreements to the value of the related exports, illustrates well the increasingly demanding nature of defence offset policies.

Fig. 1 is based on the data, collected by the United States Department of Commerce (US DoC), on offset agreements concluded by US firms during the period 1993–2011.[4] The reports on 'Offsets in Defense Trade' published by the US DoC represent the only source of aggregated data on offsets available. Their coverage is by definition limited to agreements and

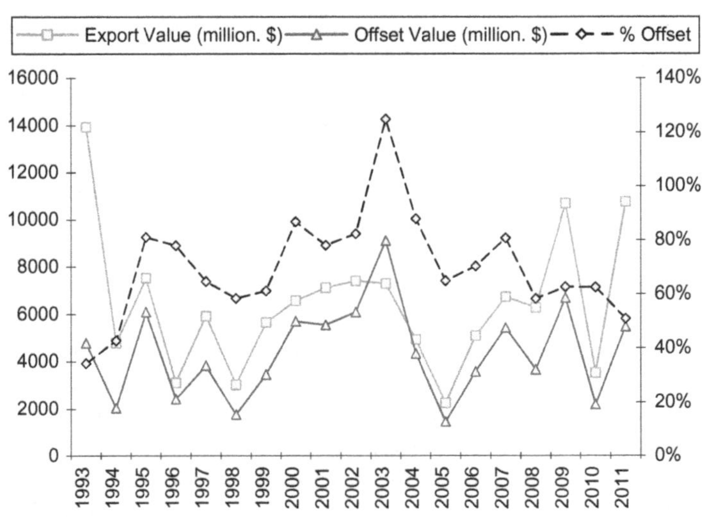

Fig. 1. Offset Agreements by US Defence Companies (USD million). *Source:* US DoC (2013).

transactions operated by US firms. Considering however that the United States is the largest exporter of defence equipment, the data set provides a good illustration of the global trends.

During 1993–2011 US firms entered in 830 offset related defence export contracts. The total value of the offset agreements concluded during the period was of approximately 83,73 USD billion. From Fig. 1 it is visible that the offset percentage[5] was following an increasing trend between 1993 and 2003. In 2003 it culminated at an historical high of almost 125% on average. Following the 2003 record, the offset percentage fell down and over the last four years covered by the data set available it varied between 51% and 63%. It is important to note the substantial instability of the series presented in Fig. 1: individual large agreements can influence the values for the years where they are concluded.

Offset-Relevant Regulation at the International Level: The GATT/WTO

At the international level the main body of legal rules relevant for offsets can be found at the level of the regulations of the World Trade Organisation (WTO) and its predecessor the General Agreement on Tariffs and Trade (GATT).

It is first worth noting that offsets can be approached as a specific form of countertrade. As such they are covered by the contributions analysing the compatibility of the latter with the WTO/GATT rules on international trade. In an article published in 1986,[6] Michael R. Czinkota and Anne Talbot reached the conclusion that countertrade contradicts several basic principles as well as some specific provisions of the GATT. In particular, the authors argued that countertrade is inconsistent with the principle of multilateralism and with the aim at reducing trade distortions. The former of those principles is breached by the fact that countertrade tends to favour bilateral commercial relations at the expenses of open multilateral and non-discriminatory trade. Through the additional costs that it introduces and by potentially supporting inefficient producers countertrade equally contravenes the latter principle. The authors also found countertrade to be in contradiction with the principles of transparency, consultation and compensation.

The GATT's principles are at the heart of the process of liberalisation of international trade initiated in the second half of the 1940s. The main objective of the latter is to reduce or suppress existing protectionist barriers and create an increasingly open world economy. Free trade should generally contribute in improving global welfare by allowing countries to specialise and an efficient international division of labour to emerge, thus avoiding costly duplications and the artificial survival of economically inefficient industries in some countries. In this regard offsets can be considered as a specific type of protectionist measure that contributes in maintaining inefficient enterprises in the market through the obligations imposed to foreign contractors. Offsets, as countertrade in general, also allow for the possibility to operate subsidies in a less transparent way thus reducing the risks of detection (Banks, 1985).

Czinkota and Talbot (1986) also argue that countertrade is likely to contravene some specific provisions of the GATT: the most-favoured-nation and non-discrimination clauses defined in Article I of the GATT agreement, the national treatment principle of Article III, the prohibition of quantitative restrictions of article XI, the subsidy code and Article XVII regarding state trading enterprises.

Offsets equally breach the provisions of some more recent agreements that form part of the WTO regulatory system. Such an example is the Agreement on Trade-Related Investment Measures (TRIMs) prohibiting all such measures contravening to the principles of national treatment and the prohibition of quantitative restrictions.[7]

Another piece of WTO legislation even introduces an explicit prohibition of offsets. Article XVI (1) of the Government Procurement Agreement (GPA) reads as follows: 'Entities shall not, in the qualification and selection of suppliers, products or services, or in the evaluation of tenders and award of contracts, impose, seek or consider offsets'.[8]

National security exception clauses have however been systematically incorporated in the GATT/WTO Agreements, thus limiting the application of the prohibition of offsets in the defence sector. The GPA is no exception, as article XXIII provides for the above-mentioned exemption.[9] It is worth noting that offsets are also common in some civil sectors such as commercial aircraft or large energy plants, but until now no case has been made attempting to enforce the ban of offsets.

Defence Offsets in the European Union

Europe has traditionally been at the avant-garde of offsets requirements. The US DoC reports frequently emphasise this fact by demonstrating that the conditions required by the European countries are on average more demanding than those imposed in the rest of the world. Over the

1993–2006 period the offset percentage for agreements concluded with European countries was of 97.74% while for agreements with non-European countries it was of only 46.73%.[10]

The only estimation of offset agreement's value and offset percentage in Europe is provided in a study on offsets ordered by the European Defence Agency (EDA).[11] On the basis of data covering the 2000–2006 period it was estimated that offset agreements represented approximately 5,65 EUR billion per year.[12] The average value of the offset percentage was estimated at 135%, with the values for individual countries varying between 72% and 237%.

Until recently, almost all European Union (EU) Member States possessed some form of offset policy. The CTO's *Offset Guidelines Quarterly Bulletin* editions of 2005–2006 listed information on the offsets policies and practices of 23 out of the current 27 EU Member States. Even large arms exporting countries such as France and Germany, which are often wrongly perceived as not accepting offsets 'as a matter of policy',[13] were ranked in the Top 20 countries with regard to the value of offset transactions executed by US defence firms during the period 1993–2006.[14] The United Kingdom systematically occupied the first rank with offset transactions totalling more than 7 USD billion over the period. In the Top 20, 13 countries were European, 12 of them EU Member States, the remaining one being Switzerland.[15]

Eastern European countries also progressively found a place amongst the largest offset demanders. At the end of the Cold War these countries were generally characterised by the presence of comparatively large national defence industries, mainly specialised in the production of defence equipment of Soviet origin. As a consequence of the collapse of internal defence spending and of the loss of their traditional markets, these industries quickly faced important difficulties. Some countries, such as Poland, reacted by relying on offsets in order to support and modernise their defence sector. In 2003 Poland decided to purchase 48 F-16 combat aircraft from Lockheed Martin. The deal included a very stringent offset agreement that was valued at over 6 USD billion for a core contract of approximately 3.5 USD billion (Markowski & Hall, 2004[16]). The resulting offset transactions projected Poland at the 4th place in the above-mentioned Top 20 countries with regard to the value of offset transactions executed by US firms. Another important offset agreement was concluded by Poland in relation to its purchase of armoured vehicles from the Finish enterprise Patria.

With regard to regulation, it may seem that the offset-relevant regulatory framework in the EU is quite similar to the one that characterises the GATT/WTO. The European Commission has traditionally been of the opinion that offsets are contrary to some of the basic principles of EU primary law, in particular those pertaining to the free movement of goods and services. As in the case of the WTO, the most relevant consequence of these principles is the ban of quantitative restrictions on trade. It is in this sense that offsets appear as being particularly problematic. The use of offsets can also be considered contrary to the principles of non-discrimination (on basis of nationality), equal treatment and transparency.[17]

Additionally, offsets infringe specific rules concerning public procurement that are defined in Directive 2004/18/EC. Above all, including offsets amongst the contract award criteria is not permitted even if the contract is awarded on the basis of 'the most economically advantageous tender'[18] criterion.

The Commission's approach has favoured an open and transparent public procurement process for defence equipment in view of facilitating the emergence of a more integrated European Defence Equipment Market. Member States, in particular those with large defence industries, have however been widely using their defence procurement policies in order to support national manufacturers. In doing so, they have relied on a wide interpretation Article 346 of the Treaty on the Functioning of the EU (TFEU).[19]

Defence procurement in the EU has therefore escaped, to a large extent and over several decades, the need to respect the above-mentioned principles and rules thanks to the extensive reliance on the exemption introduced by Article 346 TFEU. The latter have been traditionally considered by EU Member States as a general exemption that allows them to almost completely isolate the defence procurement process from the EU internal market regulation framework. One particular consequence of that was the widespread of offsets in intra-European defence trade. However, in contrast with the situation at the WTO level, recent evolutions in the European context may lead to changes with potentially substantial consequences for the defence sector in Europe.

REGULATING OFFSETS IN THE EU AND THE INTEGRATION OF THE EUROPEAN DTIB

The Challenge of Tackling Defence Offsets

One can wonder whether the wide interpretation of the provisions of Article 346 TFUE on which the EU Member States usually rely finds support in the ECJ jurisprudence and in the Commission's case practice. However, progress on this point was slowed down by the lack of rules, within the general regime on defence procurement in place in the EU, taking into account in an appropriate manner the specificities of defence procurement. The additional initiatives introduced in the framework of the Regime on defence procurement of the EDA have also produced limited effects, in particular because of their non-legally binding nature.

In 2006 the Commission, recalling the ECJ's case law, published an 'Interpretative Communication on the application of Article 296 (now 346) of the Treaty in the field of defence procurement'[20] stating clearly that the derogation provided for in article 346 TFUE does not lend itself to a wide interpretation. The need for relying on this derogation has to be demonstrated by the Member States on a case-by-case basis with a justification provided about the fact that an essential security interest is at stake (industrial and economic interest cannot serve as justification). It is also necessary to establish the connection between this essential security interest and the specific public procurement and justify the necessity of the non-application of the specific EU public procurement rules in order to protect the relevant essential security interest.

The Commission has however not immediately enforced the above-described interpretation in the field of public procurement. As mentioned before, the main reason for this is to be found in the fact that it was widely acknowledged that the general public procurement regime in the EU was not fully adapted to the specificities of the defence sector. As long as adequate rules for defence procurement were not introduced, it was difficult to enforce a stricter interpretation of the use of the national security derogations.

The difficulty in tackling defence offsets through regulatory means is however far from being a specifically European problem. In the US an intense debate regarding offsets has been present for decades. Despite heavy pressure to reduce offsets granted in defence sales to foreign countries, pressure originating from groups such as labour unions and subcontracting enterprises and usually channelled through the Congress, the US Administration's policy has been surprisingly stable.

While officially ascertaining the trade distortive and economically inefficient nature of offsets, the US policy has remained loyal to the benign neglect approach already defined in the Duncan Memorandum of 1978[21] as a result of which the US government does not participate in offsets but leaves them under the responsibility of the US defence industry. The strong opposition from the latter has been the main reason why almost none of the frequently proposed anti-offset legislations has ever been adopted, the fear being that any reduction in the industry's flexibility to offer offsets could result in losing export contracts to foreign competitors. In this respect it can be mentioned that each issue of the US DoC report on offsets contains a quantitative analysis of the effects of offsets on the US defence industrial base and defence preparedness. The reports have always reached the conclusion that the global effect of offsets on the US defence industry is positive because the work and jobs that they maintain through the exports realised outweigh those that they transfer abroad.[22]

A pragmatic approach to offsets, similar in some aspects to the one of the US defence industry, is also present in the European context. A policy paper on offsets published in 2001 by the European Defence Industries Groups (EDIG, 2001) advanced that within Europe, in the absence of a truly integrated defence equipment market, offsets can be an effective tool for opening the industry's supply chains to competition. Therefore, as long as the market has not reached the necessary level of integration, offsets will have a role to play that cannot be considered as producing purely negative consequences on the industry's structure.

In the recent years however, the increasing intensity of offset requirements, in particular as expressed in the form of offset percentages, started raising some worries at the levels of both industry and European institutions. The offset percentages put forward in some deals could lead to the fear that an offset race was ongoing.

A first attempt to regulate offsets in the EU was made in the framework of the EDA: on 1 July 2009 the Agency introduced a Code of conduct on offsets as an integral part its Regime on defence procurement. In addition to the implementation of transparency obligations requiring participating Member States to publish information on their offset policies on the Offset Portal available at the Agency's Internet site, the Code's most important feature is the introduction of a 100% cap on the offset percentage that can be imposed. The code also emphasises that the implementation of offsets should contribute to the emergence of capability driven, competent and competitive European DTIB, by favouring centres of excellence and avoiding unnecessary duplications.

The EDA's Code of conduct on offsets is however a voluntary and non-legally binding document and its real influence still needs to be demonstrated. Another legal instrument is expected to have much more stringent and far reaching consequences on the European defence equipment market in general and on offsets in particular: Directive 2009/81/EC, adopted during the summer of 2009, provides for the first time at the EU level a common set of rules regarding public procurement in the defence and security sectors.

The Introduction of Directive 2009/81/EC and its Consequences

In December 2007, the European Commission launched an initiative aimed at tackling the weaknesses caused by the fragmentation of the European defence market and known as the 'Defence package'. In an addition to a general Communication on A Strategy for a Stronger and More Competitive European defence Industry[23] and of a Directive on Intra-EU Transfers of Defence-related Goods[24] aimed at reducing the obstacles for intra-EU trade of defence products, the Defence package introduced for the first time a proposal for a common set of rules at the EU level regulating the public procurement of defence and security related products and services.

As previously mentioned, the lack of public procurement rules fitting the specific needs of the defence sector has been one of the main reasons why defence procurement has previously been subject only to a limited control at the EU level. This gap was finally bridged by the adoption in 2009 of Directive 2009/81/EC as part of the above-mentioned 'Defence package'. The latter's provisions are defined in order to allow tackling issues of particular importance in the defence procurement process such as security of supply and security of information.

Following the transposition of the Directive, the procedures it provides for 'should be considered as the standard procedures for defence and sensitive security procurement',[25] the recourse to Article 346 TFEU being thus limited to exceptional cases. While offsets are not mentioned in the Directive, its introduction will have very important consequences in that respect. Offsets will be above all impacted by the general shift of defence procurement from an Article 346 TFUE base to the new regime established by Directive 2009/81/EC. Additionally, the presence of the Directive will also enable the Commission to be more active in enforcing the strict interpretation of the exemption of Article 346 TFEU which has further consequences with regard to offsets.

The European Commission generally shares the wide agreement existing on the fact that offsets can potentially result in trade distortions, in particular by securing business for inefficient suppliers located in importing countries and which would have not obtained such opportunities without offsets. The latter would therefore contribute in creating distortions of the international division of labour, in maintaining inefficient producers, in avoiding necessary restructuring, and in upholding large over-capacities in the sector.

Taking into account that offsets can be contrary to the basic principles of the Treaty, the Directive can neither authorise them, nor regulate them.[26] The transposition of the Directive will therefore require EU Member States to review their national legislation and to suppress any form of mandatory offset policy. The new legal framework equally rules out the use of offsets as a contract award condition or as a criterion for the selection of the successful tenderer.

It is also expected that the European Commission will exert a higher degree of scrutiny on contracts awarded under Article 346 TFEU, which will make the recourse to offsets difficult even when the Directive does not apply. As mentioned above, using Article 346 TFEU requires a case-by-case justification of the necessity to impose requirements that may be contrary to EU law in order to protect an essential security interest. An important precision is that the justification should be provided for each specific requirement. Therefore, the fact that a specific contract will be awarded under Article 346 TFEU does not mean that offsets could be attached to this contract.

For the use of offsets to be possible, i.e. for them to fall under the exemption provided under Article 346 TFEU, Member States will need to demonstrate that offsets themselves are necessary for the protection of a specific essential security interest. Providing such a justification will generally be impossible for civil offsets, which have the additional flaw of being in contradiction with the provisions of Article 346 TFEU (1) (b) requiring that the conditions of competition for products 'which are not intended for specifically military purposes' should not be adversely affected. As far as defence offsets are concerned, while a justification remains possible at the theoretical level, in practice the need for offsets in order to defend a specific essential security interest will be difficult to demonstrate especially if we consider that the Directive incorporates specific provisions aiming at the protection of Members States' security of supply and security of information requirements.

A possible exception may exist in situations regarding relations with suppliers located outside the EU. In this particular case it is not impossible that some specific requirements could be justified on grounds of security of supply when the latter is spelled out at the EU level. In general rule however, the effective enforcement of the principles described in this section should result in a substantial contraction of offsets, or at least of their visibility and formal nature. This represents a major shift for the defence market in the European Union, the potential consequences of which are treated in the next section.

Offsets: A Driver of Integration or Disintegration of the DTIB?

Considering the figures quoted in section 'The Challenge of Tackling Defence Offsets' above it is obvious that offsets are a very important factor for the defence sector. It would be unconceivable to pretend that several decades of intensive offset policies have not produced any effect on the defence industry's organisation, internationalisation and efficiency.

The nature of the influence of offsets on the defence market and industry remains however ambiguous. First, a negative preconception is clearly present, sometimes impacting on the impartiality of the analysis performed. Second, discussions have been seriously plagued by diverging definitions (See also Hall & Markowski, 1994), the variety of forms that offsets can take and the resulting lack of a truly unified analytical framework. Third, the number of reliable studies available is more than limited, mainly because of the highly sensitive nature of offsets and the resulting lack of available data.

While everybody seems to agree on the potential distortions that offsets may create, one of the major shortcomings of the economic literature of offsets is to be found in the fact that a vast number of authors are already satisfied with this level of analysis. They remain thus unwilling to consider that offsets may sometimes produce positive effects or, more importantly, that they could be approached as an indicator for some crucial deficiencies of the defence equipment market rather than as an independent phenomenon.

Ianakiev (2005) demonstrated that the introduction of offsets can have powerful consequences on the exporting system integrator's incentives to search for partners in the importer's economy and to establish industrial relations with them. The incentives resulting from the conditioning of the core contract to the establishment of industrial connections with local companies can be strong enough to push the system integrator to start working with local suppliers that are less efficient than its existing partners as long as the extra costs incurred are not offset by the profits of the core contract. This effect is also reinforced by the capacity of exporting firms to successfully transfer most of the offset-induced extra costs to the importing government by integrating them ex ante in the proposed price. This confirms the trade-distorting potential of offsets.

However, if substantial supplier search and switching costs exist, spontaneous search by system integrators may be rather limited in the absence of offsets. Ianakiev (2005) shows that in such a situation, a possible outcome of the use of offsets could be the discovery of unknown suppliers, more efficient than the incumbents or at least efficient enough so that they could become second source (thus potentially contributing positively in terms of security of supply) and with which a long-term partnership could be established. This theoretical result confirms findings present in several surveys and in a number of case studies on offsets showing that in some occasions more efficient suppliers are discovered,[27] this leading to an improvement of the supply-chain efficiency in terms of costs or technological performance.

If we consider that the system integrator's supply chains have initially been set up in sheltered national markets, the joint effect of policy induced protection and supplier search and switching costs may substantially reinforce the position of incumbent suppliers (Ianakiev & Mladenov, 2008). This is especially true for an industry in which price competition has not been particularly acute. It is even possible to demonstrate that full liberalisation and suppression of protectionist measures may have only a limited effect on the international allocation of activities provided that supplier search and switching costs are perceived as being high enough to inhibit the exploration for new partnership opportunities. In the presence of supplier search and switching costs, protection results in a hysteresis effect even after the removal of protectionist regulation.

The industrial structure that previous protection has induced can effectively survive the latter's disappearance. Potentially efficient suppliers can therefore be pushed out of business because they are unable to contest the incumbents' positions. In this context offsets can also be a driver for integration in the EU defence equipment market by making it possible to overcome the barriers preventing the spontaneous emergence of trans-border industrial relations.

The duality of offset's impact on the integration process in the European defence equipment market has important consequences with respect to the recent evolution in the EU regulation framework. The reduction of offsets that should result from the evolution of the regulation framework will of course limit the risks of offset-induced distortions of the defence equipment market. It is however also possible that it may reduce the ability of some potentially efficient firms to access the system integrator's supply chains. Most vulnerable in this respect will be small and medium size enterprises (SMEs) and enterprises located in small and medium size countries that do not possess large system integrators.

The main resulting risk is the one of a possible decrease in the degree of contestability of the incumbent supplier's positions through the suppression of a powerful instrument for forcing access to the system integrator's supply chains. It will therefore be crucial to accompany the efforts to restrict the use of offsets by strong initiatives aimed at ensuring that a level playing field exists not only at the level of public procurement but also at the supply chains one.

It is particularly important to incentivise system integrators to improve their supply-chain management practices and to increasingly favour open competition at the subcontracting level. Directive 2009/81/EC already incorporates dispositions intended to enhance competition at the subcontracting level. Article 21 and Title III of the Directive provide in particular for the possibility for the contracting authority to require the successful tenderer to subcontract a share, not exceeding 30%, of the contract. In contrast with offsets, the selection of the subcontractors has to be based on a competitive procedure opened to all interested European enterprises without any form of discrimination on basis of nationality. It is however uncertain whether and to what extent Member States will make use of the Directive's subcontracting provisions.

Initiatives aimed at reducing the informational barriers and deficiencies can usefully complement the above-mentioned efforts to promote open selection procedures at the supply-chain level. Centralised publication of subcontracting opportunities can be a very useful tool in this respect. Unfortunately, the past experience with the dedicated part of the EDA's Electronic Bulletin Board, established in the framework of the Code of Best Practice in the Supply Chain, has not been fully satisfactory. International industry seminars, better focused on the needs of system integrators, in which a sizeable presence of SMEs is ensured can also contribute positively in this respect.

CONCLUSION

The wide presence of offsets represented in the past few decades one of the most prominent characteristics of international defence trade. As described above, Europe has traditionally been a leader in this respect. The fact that the European Union has also taken the lead in attempting to put the phenomena under control and to improve enforcement of the rules prohibiting these practices may appear surprising, especially if we consider the long lasting and often vocal criticism of offsets coming from the other side of the Atlantic. However, taking in consideration the high value of intra-European arms trade, there is a clear logic in limiting the sources of distortions while trying to build an increasingly efficient, competitive and genuinely integrated European defence equipment market.

If this ambition is to have real chances of success it will however be necessary to better understand the imperfections currently affecting the European defence equipment market and to acknowledge the fact that offsets may also have played, in some circumstances, a positive role in the process of integration of the European defence equipment market. Creating open competition and easy access at the level of the large system integrators' supply chains is one major challenge that lies ahead. If the use of offsets is to be restricted in the future, new instruments should be quickly put in place in order to ensure that this evolution does not result in driving out of the market potentially efficient suppliers only because of the obstacles that they are facing with regard to the access to the system integrator's supply chains.

NOTES

1. Udis and Maskus (1991).
2. SIPRI (2009).
3. Directive 2009/81/EC of the European Parliament and of the Council of 13 July 2009 on the coordination of procedures for the award of certain works contracts, supply contracts and service contracts by contracting authorities or entities in the fields of defence and security, and amending Directives 2004/17/EC and 2004/18/EC (2009).
4. United States Department of Commerce (2013).
5. The offset percentage is the ratio of the value of offset agreements to the value of the related exports. It is the most adequate measure of the level of offset requirements, offsets agreements value being very sensitive to year by year changes in the value of related exports sales.
6. Czinkota and Talbot (1986).
7. The illustrative list provided in Annex to the TRIMs Agreement states the following: 'TRIMs that are inconsistent with the obligation of national treatment provided for in paragraph 4 of Article III of GATT 1994 include those which are mandatory or enforceable under domestic law or under administrative rulings, or compliance with which is necessary to obtain an advantage, and which require: ... (a) the purchase or use by an enterprise of products of domestic origin or from any domestic source, whether specified in terms of particular products, in terms of volume or value of products, or in terms of a proportion of volume or value of its local production ...'.

8. In the GPA offsets in government procurement are defined as '... measures used to encourage local development or improve the balance-of payments accounts by means of domestic content, licensing of technology, investment requirements, counter-trade or similar requirements'.

9. Article XXIII (1) of the GPA provides that 'Nothing in this Agreement shall be construed to prevent any Party from taking any action or not disclosing any information which it considers necessary for the protection of its essential security interests relating to the procurement of arms, ammunition or war materials, or to procurement indispensable for national security or for national defence purposes'.

10. United States Department of Commerce (2007). In the more recent issues of its report on *Offsets in Defence Trade*, the DoC no longer provides information on the regional distribution of offset agreements.

11. FOI and SCS (2007).

12. The study covers 24 EDA's participating Member States. Denmark, Bulgaria and Romania are not covered. The first country because it decided not to participate in the EDA, the last two because the study was conducted before their accession to the EU and the EDA.

13. FOI and SCS (2007). op. cit., 4.

14. US DoC (2007). op. cit., 5–3.

15. Turkey, which is a candidate country for EU membership also ranked in the top 20. However, according to the classification adopted in the US DoC reports on offsets, Turkey is accounted for in the Middle East and Africa group.

16. Markowski and Hall (2004).

17. Directorate General Internal Markets and Services, European Commission (2010a).

18. *Ibid*, p. 5.

19. Article 346 TFEU (former Article 296 of the EC Treaty) reads as follows:

'1. The provisions of the Treaties shall not preclude the application of the following rules:

(a) no Member State shall be obliged to supply information the disclosure of which it considers contrary to the essential interests of its security;

(b) any Member State may take such measures as it considers necessary for the protection of the essential interests of its security which are connected with the production of or trade in arms, munitions and war material; such measures shall not adversely affect the conditions of competition in the internal market regarding products which are not intended for specifically military purposes ...'.

20. European Commission (2006).

21. In May 1978, US Deputy Secretary of Defence Charles Duncan signed a document stating formally that the US administration will not, as a general rule, enter in offset agreements, thus leaving the full responsibility in that matter to the industry. The Duncan Memorandum was to a large extent a result of the difficulties that the US Department of Defence experienced in the 'Peace Alps I' deal (sale of F-5 combat aircraft to Switzerland) where it had guaranteed the offset obligations contracted by the exporting US firms.

22. The analysis performed in the US DoC reports is however quite simplistic and relies on a number of assumptions that are both questionable and prone to substantially influencing the outcome.

23. Communication from the Commission to the European Parliament, the Council, the European Economic and Social Committee and the Committee of the Regions – A strategy for a stronger and more competitive European defence industry (2007).

24. Directive 2009/43/EC of the European Parliament and of the Council of 6 May 2009 simplifying terms and conditions of transfers of defence-related products within the Community (2009).

25. Directorate General Internal Markets and Services, European Commission (2010b).

26. Directorate General Internal Markets and Services, European Commission (2010a).

27. A survey by Martin and Hartley (1995) on the experience of British exporting firms with offset requirements provides the following result: 'In six of the 11 offset sales the respondents said that the offset obligation had led to the discovery of new, lower cost, sources of supply and in all six cases the intention was to continue to do business with new sources once the offset obligation had been fulfilled'. The same possibility is confirmed by the study on offsets order by the EDA. Another example is provided by Redlich and Miscavage (1996): 'Members of the client company recognised that by acting as a catalyst to bring representatives from numerous parts of the corporation to Israel, the need to develop offsets opened up opportunities that otherwise would have gone unnoticed, or still worse, gone to the competition. Of the 100 potential projects, over 30 were rated as top priority initiatives. Several were signed and executed towards the end of 1993, prior to any selection of the plane or the engine, due to their enormous relevancy and contribution to the needs of the company'. (Redlich & Miscavage, 1996 in Martin, 1996, p. 404).

REFERENCES

Banks, G. (1985). Constrained markets, 'surplus' commodities and International barter. *Kyklos*, *38*(2), 249–267.

Brauer, J., & Dunne, J. P. (Eds.). (2004). *Arms trade and economic development*. London: Routledge.

Czinkota, M. R., & Talbot, A. (1986). GATT regulation of countertrade: Issues and prospects. *The International Trade Journal*, *1*(2), 155–174.

Directorate General Internal Markets and Services, European Commission. (2010a). *Directive 2009/81/EC on the award of contracts in the fields of defence and security – Guidance Note* – Field of application. Retrieved from http://ec.europa.eu/internal_market/publicprocurement/docs/defence/guide-offsets_en.pdf

Directorate General Internal Markets and Services, European Commission. (2010b). *Directive 2009/81/EC on the award of contracts in the fields of defence and security – Guidance*

Note – Offsets. Retrieved from http://ec.europa.eu/internal_market/publicprocurement/docs/defence/guide-scope_en.pdf

Directive 2009/81/EC of the European Parliament and of the Council of 13 July 2009 on the coordination of procedures for the award of certain works contracts, supply contracts and service contracts by contracting authorities or entities in the fields of defence and security, and amending Directives 2004/17/EC and 2004/18/EC. (2009). *Official Journal of the European Union*, 52(L 216), 76–136.

Directive 2009/43/EC of the European Parliament and of the Council of 6 May 2009 simplifying terms and conditions of transfers of defence-related products within the Community. (2009). *Official Journal of the European Union*, 52(L 146), 1–36.

EDIG. (2001). *EDIG policy paper on offsets*. EPP/00/18. Brussels.

European Commission. (2006). *Interpretative communication on the application of Article 296 of the Treaty in the field of defence procurement*. Retrieved from http://eur-lex.europa.eu/LexUriServ/LexUriServ.do?uri=COM:2006:0779:FIN:EN:PDF

European Defence Industry. (2007). European Commission.

FOI, & SCS. (2007). *Study on the effects of offsets on the development of a European defence industry and market*.

Hall, P., & Markowski, S. (1994). On the normality and abnormality of offsets obligations. *Defence and Peace Economics*, *5*(3), 173–188.

Ianakiev, G. (2005). How do offset policies affect the international division of labour? The case of the defence industry. Paper presented at the Ninth Annual Conference on Economics and Security.

Ianakiev, G., & Mladenov, N. (2008). Offset policies in defence procurement: Lessons for the European defence equipment market. *Défense Nationale et Sécurité Collective, Hors série*, 185–194.

Kirkpatrick, D. (1995). The rising cost of defence equipment – The reasons and the results. *Defence and Peace Economics*, *6*, 263–288.

Kirkpatrick, D. (2004). Trends in the cost of weapon systems and the consequences. *Defence and Peace Economics*, *15*(3), 259–273.

Markowski, S., & Hall, P. (2004). The defense industry in Poland: An offset-based revival? In J. Brauer & J. P. Dunne (Eds.), *Arms trade and economic development* (pp. 172–186). London: Routledge.

Martin, S. (Ed.). (1996). *The economics of offsets: Defence procurement and countertrade*. Amsterdam: Harwood Academic Publishers.

Martin, S., & Hartley, K. (1995). Defence equipment, exports and offsets: The UK experience. *Defence Analysis*, *11*(1), 21–30.

Pugh, P. (1986). *The cost of seapower*. London: Conway Maritime Press.

Pugh, P. (1993). The procurement nexus. *Defence Economics*, *4*, 179–194.

SIPRI. (2009). SIPRI Yearbook 2009. Oxford: Oxford University Press.

Udis, B., & Maskus, K. (1991). Offsets as industrial policy: Lessons from aerospace. *Defence Economics*, *2*, 151–164.

United States Department of Commerce. (2007). *Offsets in Defense Trade* (No. 12).

United States Department of Commerce. (2013). *Offsets in Defense Trade* (No. 17).

CONCLUSION

Contrary to what one could expect, defence is a changing world. Its boundaries and features have been moving in the past decades. While the realm of defence was considered as quite stable through the Cold War, the past three decades are characterised by many transformations that have been impacting, at the same time, defence missions and its economic, geographical and social footprint.

Through topics explored here, this book aimed to provide keys to understand what the realm of defence is in the early twenty-first century or what it could become in the forthcoming years. When we look closely, it appears that these transformations are not achieved. Geostrategic, social and economic parameters are bound to change the environment of defence and its missions, and many adjustments are yet to be done.

Indeed, there are several lessons to be learned from recent military operations. Iraq and Afghanistan changed the role of armed forces and their approach of operations in terms of objectives and means. Similarly, missions in Libya or Mali and more globally peacekeeping operations opened new dimensions regarding the use of military power.

Additionally, armed forces benefitted from quite large resources during the post-Cold War. Financial and budgetary crises in the post-2007 era deeply modify the economic landscape even for armed forces. While security and geostrategic issues are still critical, many nations can no longer afford to dedicate a substantial share of their GDP to defence. Even though emerging countries dedicate an increasing share of their wealth to security issues, these countries cannot afford advanced military systems and their efforts remain still well below what industrial countries invest in their defence.

The only means to sustain military operations is to transform again the relationship between defence and the whole society. Here, one can expect disruptive changes. This concerns to a large extend relations between armed forces and industry, but it is also likely that civil–military relations have to be revisited.

Beyond the use of forces, this raises the question of the non-traditional approach to defence and security to overcome budgetary constraints.

The effectiveness of defence can represent a new field of research. Indeed, Two dimensions are worth being mentioned.

First, defence efforts appear to be engaged into an unsustainable trend. Arms programmes certainly provide the best example. Even the largest countries appear unable to acquire all systems that their armed forces could expect. In 1983, Norman Augustine predicted that if the long-term increase of per-unit cost of major systems was not interrupted, the whole defence budget of the United States would only be able to buy a single aircraft in 2054. It seems that such prediction could become true even sooner than predicted when one analyses the per-unit cost of advanced systems like F-35 fighter aircraft or Nimrod MRA4 maritime patrol and attack aircraft.

Second, one can wonder what kind of output defence effort should produce. Indeed, defence remains based on a national perspective, conceived in the nineteenth century when nation-states were looking from defence means against potential foes. Nevertheless, stakes have changed in the twentieth century and even more on recent years. While protecting national sovereignty, population and territory remain critical dimensions, armed forces are more and more committed to peacekeeping and preserving international security. This new approach leads to a collective perspective of peace and security, with profound impacts on the ways armed forces conceive their missions and means.

These two dimensions show that there are many questions to deal with in order to truly understand the realm of defence and its evolutions. Defence deserves that academics investigate this field of research.

Defence appears as a complex field of research, since its boundaries are not easy to define and there are many political interferences that can modify usual assumptions in economics or social sciences, particularly concerning criteria of choice and cost-benefit analysis. Nevertheless, defence absorbs a large share of public spending and resources; therefore, this field of research cannot be neglected. Moreover, its economic and social footprints are numerous even today, as underlined through this book. It seems then necessary to explore the realm of defence and set up tools to understand their multiple implications.

Even though many fruitful research projects were carried on in the 1980s and 1990s, defence has become less attractive by the late 1990s. It has not received as much attention as one could expect despite several changes and challenges that have occurred since the end of the Cold War. Nevertheless, as underlined throughout this book, many questions are still to be addressed. Defence remains an uncharted field to a large extent.

Moreover economics and social sciences themselves are characterised by any changes since the 1980s. One can expect to mobilise many of these new theoretical and analytical approaches to better understand the realm of defence as well as its evolving boundaries. Indeed, there is a consistent literature in defence economics, but most of these papers and books appear quite old and they do not leverage on scientific advancements over the past three decades. Existing literature in defence economics and social sciences can and should be revisited.

This book tries to open new fields of research regarding the economic and social impacts of defence. Indeed, this is a partial work, since it focuses on selected dimensions. The questions apprehended through these thirteen chapters are nevertheless key features of today's defence. Additional dimensions would deserve further research.

As this book underlines, the realm of defence is still characterised by several terrae incognitae, some of them that have been identified for a long time, others that have just appeared or are emerging. Even though defence can be considered as a marginal topic for economics and social sciences, it therefore deserves that academics and scientists explore its dimensions to guide decision-makers and help citizens understand what is at stake.